# *Turn on the Words!*

# *Turn on the Words!*

## Deaf Audiences, Captions, and the Long Struggle for Access

**Harry G. Lang**

Gallaudet University Press
Washington, DC

Gallaudet University Press
gupress.gallaudet.edu

Gallaudet University Press is located on the traditional territories of Nacotchtank and Piscataway.


Printed in the United States of America

ISBN 978-1-944838-84-3 (paperback)
ISBN 978-1-944838-85-0 (ebook)

*Library of Congress Cataloging-in-Publication Data*

Names: Lang, Harry G., author.
Title: Turn on the words! : deaf audiences, captions, and the long struggle for access / Harry G. Lang.
Description: Washington, DC: Gallaudet University Press, [2021]. | Includes bibliographical references and index. | Summary: "The evolution of Captioned Films for the Deaf (CFD) program is a primary focus of the book. Topics such as movie theater captioning and closed captioned television are also featured, as well as the role played by deaf community activists to improve access via captions"— Provided by publisher.
Identifiers: LCCN 2021005292 (print) | LCCN 2021005293 (ebook) | ISBN 9781944838843 (paperback) | ISBN 9781944838850 (ebook)
Subjects: LCSH: Motion pictures—Titling. | Television programs—Titling. | Closed captioning. | National Association of the Deaf. Described and Captioned Media Project.—History.
Classification: LCC TR886.9 L34 2021 (print) | LCC TR886.9 (ebook) | DDC 777/.55—dc23
LC record available at https://lccn.loc.gov/2021005292
LC ebook record available at https://lccn.loc.gov/2021005293

∞ This paper meets the requirements of ANSI/NISO Z39.48–1992 (Permanence of Paper).

Image on top of cover: Photograph by Sally Taylor, courtesy of the Paul and Sally Taylor collection at Southwest Museum of Engineering, Communications and Computation (Ed Sharpe, director). www.smecc.org. Lower cover picture: Gallaudet College students watching the captioned film *Cyrano de Bergerac*, in 1961. Image courtesy of Gallaudet University Archives.

Cover design by Eric C. Wilder.

Cover description: Top of the cover is a photo of people marching down the street during a CBS protest in Rochester, NY. Below the photo is the title and subtitle: Turn on the Words! Deaf Audiences, Captions, and the Long Struggle for Access. Below the title is a photo of students sitting in rows in a classroom with their backs to the camera. They are watching a captioned film on a projector screen. Superimposed on the classroom image is the author's name: Harry G. Lang.

## CONTENTS

# FOREWORD

Ernest E. Hairston

Why should a book on captioned films for deaf people be written? Why is it important that the story be told? Captioned movies shown in theaters nationwide and captioned television programs are now taken for granted. Originally, captioned films were produced in order that deaf and hard of hearing viewers could understand what was spoken on the screen instead of guessing or imagining what was being said. In fact, captions have always benefited a larger and varied audience, especially people learning English as a second language and patrons of noisy places such as bars and restaurants. But captions have empowered deaf and hard of hearing people. Access has led to increased success. And as Harry Lang has shown in his history of the TTY (*A Phone of Our Own*), many of the advances in both telephone technology and captioned films involved deaf people. As described in both books, deaf people should take pride in the roles they played in these technological advances.

Having lived through much of the early stages of the evolution of both telephony and captioning, I gave little thought to how important history is to education until I read Harry's books. New generations of deaf people have available a variety of captioning technology and other telecommunication devices available to them in their youth. Reading about how hard the previous generations worked to gain access to leisure time entertainment may provide them with the motivation to continue the legacy. Those who know this history may be inspired to enrich it further.

For older deaf people like me, who learned to enjoy and depend on captioned movies and captioned television the hard way, one small breakthrough at a time, we know well what it was like. But there are surprises for us, too, in this book. Many readers may not realize the extent to which the captioning movement spawned the rapid rise of visual media as valuable resources in the classroom. Even the deaf and hearing pioneers we personally knew were often too modest to share the many challenges they faced—the resistance of Hollywood studios to support open captions in theaters, the attitudes of people about captions being "distracting," and the months-long wait for a captioned film to become available are just a few examples.

What began as a small program called the Captioned Films for the Deaf to distribute 16 mm films to schools and the Deaf community has blossomed into a modern multimedia resource program that enriches educational environments, utilizing online streaming of films, YouTube, Facebook, and a website accessible anytime and anywhere.

Harry Lang has documented many aspects of Deaf history and Deaf studies in his various writings, and I am pleased that he has undertaken this challenge of describing the complex history of captioning as it pertains to films and television.

After more than forty years of federal service, all with the Department of Education (formerly the Office of Education within the Department of Health, Education, and Welfare), I often reminisce about my own role in implementing policies that helped bring access to myself and thousands of other deaf and hard of hearing people. I often have wondered if a book documenting the evolution of captioning and the roles of key players would ever be written while memory is fresh and readily available. Even though I managed the Captioned Media Program, as well as personnel preparation (teacher training) projects, research grants, and the National Theatre of the Deaf grant, I seldom thought about how the captioned films story was unfolding. This is it!

This book is thoroughly researched and will be a welcome addition to the growing collection of histories of the roles deaf and hearing people have shared together in applying technological breakthroughs to improve the quality of our lives.

## PREFACE

Where would deaf adults and deaf kids be today if captioning had not been developed for films? Few would argue with the belief that the impact of captioning has been dramatic. The power of words through captions has been very evident to me as I experienced this access over the last few decades. Captions have provided me with knowledge, amusement, and satisfaction in my life. As an educator, I have seen firsthand how they have empowered my students by providing both primary and vicarious learning experiences.

The story of how captioning came into the lives of deaf and hard of hearing people has not been told in any detail, and I welcomed the opportunity to document it. When Gilbert "Gil" L. Delgado first shared with me his ideas for the book, I was satisfied with the primary focus on the government-sponsored program Captioned Films for the Deaf (CFD), which launched captions for media. Such a book appeared to me, at first, as one both concise and focused. It would be about a small program sponsored by the federal government to help bring access to deaf people through the technology of captioning. Never did I think when I began this book that the story behind this "small" program was so far-reaching and impactful.

But while focusing on the history of CFD, I realized quickly that the program had a much broader impact. CFD was a change agent that led to many technological and sociological advances. In the early 1960s, for example, the Department of Education saw the potential of CFD's early efforts to introduce captioned films and establish a mandate to enhance the education of deaf students through greater emphasis on other visual media in classrooms around the country. Through five years of summer workshops and other projects, CFD supported this goal with research and teacher training with an emphasis on not only captioned films but also filmstrips, overheads, and other media. This, and the introduction of media to schools around the country, led to a revolution.

The films distributed by CFD under the Department of Education were originally open captioned. They addressed the educational and entertainment needs of deaf children and deaf adults. Learning about culture was an important emphasis in the public laws and amendments passed throughout the first two decades of this program.

In the late 1950s and early 1960s, there were many promises made of leveling the playing field for access to educational and entertainment films. Despite all the wonderful developments over the years with regard to opportunities to see captioned entertainment films, the "equality" and accessibility to feature films promised in the original legislation establishing the program had not been realized. Thus, again, the story of captioning in this book moved outside the realm of CFD as a government program when ongoing legal cases around the country became necessary for deaf people to gain access to movie theaters. Unwieldy closed captioning rearview mirror systems and special eyewear still frustrate moviegoers to this day, and there remains a strong preference among deaf people for open captions, those that are projected directly on the screen without special equipment. Local groups around the country, individual advocates, and policy experts are still fighting for that equality that never fully came. Even though it may not be the last frontier in this story of evolution and revolution, access to films in movie theaters, something deaf people originally had through the "silents" when this story began in the 1920s, fell through the cracks as new legislation was passed.

Then, in the 1990s, four decades after CFD began its work, the Americans with Disabilities Act (ADA) and the Individuals with Disabilities Education Act (IDEA) led to significant mandated changes that affected CFD. With the excitement over closed captioned television, Blockbuster Video stores, and other resources, the Deaf community was simultaneously losing all services to deaf adults when CFD began curtailing its offerings upon request by the Department of Education. The distribution of entertainment films was also stopped, and with that decision, the emphasis on "cultural awareness" through the program's entertainment films ceased.

More mandates came to address the needs of children with other disabilities and those for whom English was not their first language. Thus, while constantly addressing one Department of Education mandate after another and offering excellent services for K–12 education, the CFD program lost its grip on meeting deaf people's needs for feature film access, which were meant to help foster cultural awareness. This left deaf people waiting and hoping for a breakthrough (technological or otherwise) that would bring them that promised accessibility and equality.

The examination of the history of educational films reveals that teachers have played a significant role in the development and advocacy of captioning. Long before Thomas A. Edison developed the motion picture projector, teachers of deaf students recognized that visual materials could enhance instruction. They experimented with pictures, "lantern slides," charts, maps, and stereographs, and they immediately saw the potential of "moving pictures" when the technology became available. Interestingly, the history of the education of deaf children is packed with stories of daring teachers who were innovative in the classroom. And that spirit was especially true when motion pictures came out. Movies led to a new dream for teachers—a dream of additional technological breakthroughs that would make films accessible and shatter the isolation from many aspects of American culture that deaf people experienced.

As motion pictures gained in popularity as an entertainment medium, deaf and hard of hearing people of all ages were also excited about a new way to spend their leisure time, and they recognized that such "recreational" films would give them vicarious learning experiences related to the cultures of America as well as other countries. Motion pictures known as the "silents" immediately became a form of shared entertainment for hearing, hard of hearing, and deaf people. How much sharing of culture occurred, however, was quickly found to be dependent upon the *accessibility* of the films.

In 1927, the introduction of the "talkies" (films with the soundtrack added) dealt the first blow to film access. It took about three decades to develop the technology for captions to accommodate for the loss of the title cards that presented the dialogue in the silent films. In 1996, Congress passed legislation requiring closed captions on most television programs. Perhaps because of this, in the early 2000s—after about forty years of distributing entertainment films—the Department of Education determined that its captioning program should focus only on educational films and should stop all other captioning services for deaf adults. For this reason, in this book I step out of the history of CFD again—this time to examine the frustrating struggle of the American Deaf community of moviegoers to regain access to feature films through open captions, by and large their preference. I believe this is necessary in order to give the reader a sense of the continued relevance of such access. When the Department of Education asked CFD to discontinue the entertainment film distribution to the Deaf community, this went against the original mandate in 1958 to enhance more awareness of American culture.

In 1991, the Department of Education transferred the management of the captioned films program, henceforth known as the Described and Captioned Media Program (DCMP), to the National Association of the Deaf (NAD), and since then the program has commanded a nationwide network of accessible materials, including materials in Spanish. DCMP has also become the nation's leading source for accessible educational content and services for students who are blind, visually impaired, or deafblind as well as those who are deaf and hard of hearing. Today, families and school personnel with students from preschool through grade 12 can register for free access to more than 6,000 educational media titles on demand and on DVD.

Both before and after the NAD began to host the program, the organization played an integral role in efforts to achieve access through captions, including the provision of captions on television and access to movies in theaters through open captions. With simultaneous activism growing toward increased access to the telephone in the United States, many parallels could be observed in the two movements in the Deaf community.

My study of the history of captioned films for deaf people also shows how the success of CFD inspired other people and programs to develop a range of related technologies and services. That is, thanks to CFD's accomplishments in making films increasingly accessible, other people outside the program began experimenting with real-time captioning, captioning of YouTube videos and other videos on the internet, captioned telephones, and a host of other breakthroughs. Organizations such as Telecommunications for the Deaf and Hard of Hearing, Inc. (TDI) and the National Captioning Institute (NCI) have become leaders in these technological developments. When the Coalition for Movie

Captioning (CMC) was formed in the early 2000s as the NAD was restructuring its priorities, TDI joined this group, and TDI now leads a group of organizations serving deaf and hard of hearing people in filing policy comments with the Federal Communications Commission (FCC) regarding captioning issues.[1]

These and other organizations have collaborated with DCMP to enhance access for blind and visually impaired people through descriptions and related services. A thorough examination of such resources, however, is beyond the scope of this book, and I can only briefly mention some to reveal how DCMP has become part of the rich legacy of access.

On June 11, 1992, Frank B. Withrow from the Department of Education wrote to John A. Gough two years before Gough died, "When you think about it, this little program has done much to bring deaf people into the mainstream of society and to offer many minorities an opportunity to make significant contributions."[2] In effect, working over six decades to meet federal mandates and accommodate accessibility for a wide range of citizens, the story may well be titled "The Little Program That Could" after *The Little Engine That Could*. The story of CFD is about morality (the wrongness of not providing access) and imagination (the ingenuity of the people who brought captions to deaf people). The minimally staffed captioning program is like the story's small steam engine that manages to tow a train (the task of developing the captioning technology) over a mountain (attitudinal, financial, and communication obstacles). The moral of the story is that difficult problems can be solved when everyone remains optimistic and works together—passionately.

Activism by deaf consumers played a key role in increasing access and cultural awareness through technology. I previously authored what I consider a companion volume, *A Phone of Our Own: The Deaf Insurrection against Ma Bell*. The present story of the history of captioned films has remarkable parallels with the development of telephone access. As a deaf scientist myself, I am only too thrilled to study how both telephone and captioning technologies have made such a difference in the lives of deaf people. Both stories are fascinating summaries of how deaf and hearing people have collaborated to enrich lives.

The story of captioning is far from over. I look forward to a sequel to this history.

## TERMINOLOGY

Captioning is the process of transcribing and making visible the dialogue and other auditory information on videos and films in order to provide access to people who are deaf or hard of hearing. Spoken dialogue appears as printed words on a screen or monitor.

**Open captions** are visible on the screen to all sighted viewers. **Closed captions** appear only when electronic decoding circuitry is activated by the user. Real-time captions are broadcast simultaneously with the live program.

Some interpreter service agencies provide communication access real-time transcription (**CART**) captioning as a service for live performances and conference presentations. CART is a stenography-based system that requires stenography equipment. **C-Print** is a speech-to-text (captioning) technology and service developed at the National Technical Institute for the Deaf (NTID), a college of Rochester Institute of Technology (RIT). The system is being used successfully to provide communication access to individuals who are deaf or hard of hearing in many academic programs around the country. In addition to educational environments, C-Print also can be used in business and community settings, and with individuals with other disabilities, such as those with low vision or a learning disability (https://www.rit.edu/ntid/cprint/). As with CART, C-Print provides a real-time display of the spoken information and transcript (notes). C-Print captionists, however, provide a "meaning-for-meaning representation" of the spoken information versus verbatim, which is more often the aim of CART providers. In comparison, C-print is a concise and thorough translation of spoken English content (https://www.rit.edu/ntid/cprint/support/common-faq).

Audio description (also referred to as "video description" in the FCC's rules) is audio-narrated descriptions of key visual elements of a television program or film inserted into natural pauses in the program's dialogue (https://digital.gov/2014/06/30/508-accessible-videos-how-to-make-audio-descriptions/). Description provides a translation of images into spoken words. It has been found to be very helpful for people with vision disabilities.

## PROGRAM NAMES

| | |
|---|---|
| 1949 | Captioned Films for the Deaf (CFD) |
| 1955 | Captioned Films for the Deaf, Inc. (company becomes incorporated) |
| 1958 | Captioned Films for the Deaf (federal law) |
| 1984 | Captioned Films/Videos for the Deaf (CFV) |
| 1991 | Captioned Films/Videos (The words "for the Deaf" dropped from program name due to a formal request from the hard of hearing community) |
| 1998 | Captioned Media Program (CMP) |
| 2006 | Described and Captioned Media Program (DCMP) |

## ORGANIZATIONS RESPONSIBLE FOR THE WORK

| | |
|---|---|
| 1947–1958 | CFD/CFD, Inc. |
| 1958 | CFD (Under the federal government) |
| 1961 | Conference of Educational Administrators Serving the Deaf (CEASD). Captioned Films for the Deaf Distribution Center (CFDDC) |
| 1980 | Association for Education of the Deaf (AED), Special Materials Project (SMP) |
| 1984–1985 | Divided: (1) Selection and Captioning by AED/SMP; (2) Distribution by Modern Talking Picture Service (MTPS) |
| 1985–1991 | MTPS |
| 1991–1995 | (1) Selection and Captioning by National Association of the Deaf (NAD); (2) Distribution by MTPS |
| 1995–present | NAD |

# INTRODUCTION

By the time my father turned ninety years old, he had lost much of his hearing. Although suffering from dementia, he did remember the joys I had experienced in gaining access to films and television as captioning technology developed through the years. Every time I visited him in assistive care and nursing home facilities, he would become excited. He needed the same access to captions, but he had great difficulty remembering how to adjust the television. When I arrived for my visits to see him, he would often point to the TV and say, "Turn on the words!" After I helped him set up the captions, we would then enjoy films or baseball games together. I realized that this access was one more thing my dad and I shared. Writing this book brought back a flood of such memories.

As a deaf person, I have lived through the changes captioning has brought forth. In this introduction, I describe just a few that will especially show the reader unfamiliar with deafness what the world of deaf and hard of hearing people has been like over the past six decades.

I became deaf from meningitis in 1963 when I was fifteen years old. My family adapted well to my deafness, but my parents and sisters knew very little sign language. I struggled to read lips, and there was little they could do for me to improve my access to television or films. It would be decades before closed captions came to television. When I watched television shows, I would never understand the dialogue and was often unable to guess the plot of a movie or a shorter show like an episode of *Gunsmoke*. It was just no fun!

One very distinct memory I have from 1963 was the assassination of President John Fitzgerald Kennedy. I remember being glued to the television, hoping for any kind of graphics or print information that would help me understand this critical event in our nation's history.

My first experience with a captioned film was in 1963 when I was a high school student boarding at the Western Pennsylvania School for the Deaf (WPSD) in Pittsburgh. Because my parents lived only about a forty-five-minute drive from the school, I used to go home on weekends to see my family. Around this time, Captioned Films for the Deaf (CFD), recently established, was just beginning to distribute films to schools. When WPSD began to show captioned feature films on Sunday evenings, I would excitedly look forward to getting back to school each weekend. Like most schools at this time, there was little use of media in the WPSD classrooms. I still remember the titles of some of the Sunday evening films from fifty-five years ago. I sat watching the captions and was so absorbed in the comprehension they brought that I had a real sense of hope for the future. It seemed like *Operation Petticoat* made me laugh for the first time while watching a film in a very long time. During the summers, I would avoid theaters unless a film was animated like Walt Disney's *The Sword in the Stone.*

I was so focused on my schooling and preparing for college at that time that events in the Deaf community were foreign to me. CFD was established in 1958, only a few years earlier. It was about this time, in the early 1960s, that John A. Gough, the hearing director of the program, was advocating for access to films for deaf and hard of hearing people. In his efforts to be convincing, Gough would tell hearing audiences to sit before a television screen for just five minutes with the picture on and the sound off. "A picture is worth a thousand words, goes the old saying," he explained, "but the truth is that pictures without words are imperfect media of communication."[1] At the time, I certainly would have appreciated knowing that there were hearing leaders in the government, like John Gough, and leaders in the Deaf community, like Malcolm "Mac" Norwood, who were fighting for access for us in this manner. It would have given me a sense of hope. For now, I had to deal with a single recreational movie each week.

At college, I majored in physics and lapsed into a four-year spell with almost no captioning as part of my life. More and more educational films were being used in college classes in the late 1960s, but they were not captioned. Physics was not an easy major to study, but I had my heart set on it. Gallaudet College (now Gallaudet University) had accepted me, but physics was not offered as a major. I attended Bethany College in Bethany, West Virginia, where I graduated in 1969. However, interpreting had yet to become a profession, so most of my learning experiences came from reading textbooks or occasionally sitting next to a hearing friend who took notes during the films or lectures. By the time I finished my undergraduate studies, I had almost forgotten what the word *access* meant.

I particularly remember the television coverage of the Apollo 11 moon landing. I was home that historic day on July 20, 1969, just after I graduated from college. This was another memorable example of a critical event in our country's history, but now, watching as a young scientist, I found television without captions disappointing. When the American astronaut Neil Armstrong placed his foot onto the moon's surface—it was indeed a "small step for man, one giant leap for mankind." Yet what I saw was great irony. Our country's scientists could send a spaceship to the moon and back, but we couldn't put captions on television for millions of deaf people who were watching it!

The following year, I was offered a position teaching physics at NTID at RIT in upstate New York. I soon joined groups of new deaf friends who were getting together once per month to enjoy an open captioned film. My closest new friend, Bob Panara, had been involved with captioning through formal training at Gallaudet College, and he created the captions for many films that would be distributed nationally. He was a literature professor with a great sense of humor. I liked Bob's choice of movies for our local group, with such titles as *South Pacific* and *Paint Your Wagon*, but sometimes we had to wait three or four weeks or longer to see such a film.

Then, when RIT began showing some captioned films for the hundreds of deaf students, faculty, and staff at the new college (NTID) on campus, there was a spread of joy among us. One weekend in 1972, I made the terrible mistake of taking my future wife to see a captioned film titled *The Omega Man* starring Charlton Heston. It was a story about ghouls left on Earth after biological warfare. A half-century later, Bonnie still teases me about my early taste in films. She had degrees in literature and theater and thought the script just awful. She handed me a book, *The Great Gatsby*, hoping I might pursue better genres of entertainment. But for me, after decades of no film access, ANY captioned film was better than nothing!

Shortly after this, while teaching physics at NTID in the early 1970s, I saw the potential of captioned films as a learning aid for my deaf students in class. Almost no captioned movies were available in physics, so I began producing very crude captions for my own instructional films. With students having various communication skills, I would sign the demonstrations of a physics concept while the captions written on index cards were held up before the camera by a student assistant. Many of us teaching in colleges had not yet learned about the captioned films program's slowly growing library, which included science films. There was no internet yet, and captioned educational films still seemed like a distant dream. I had to make do with my own innovations in the classroom.

In the 1980s, closed captioned television began to emerge, but we had years of struggle ahead before having comfortable access to TV programming. We had to rally against CBS for not joining the

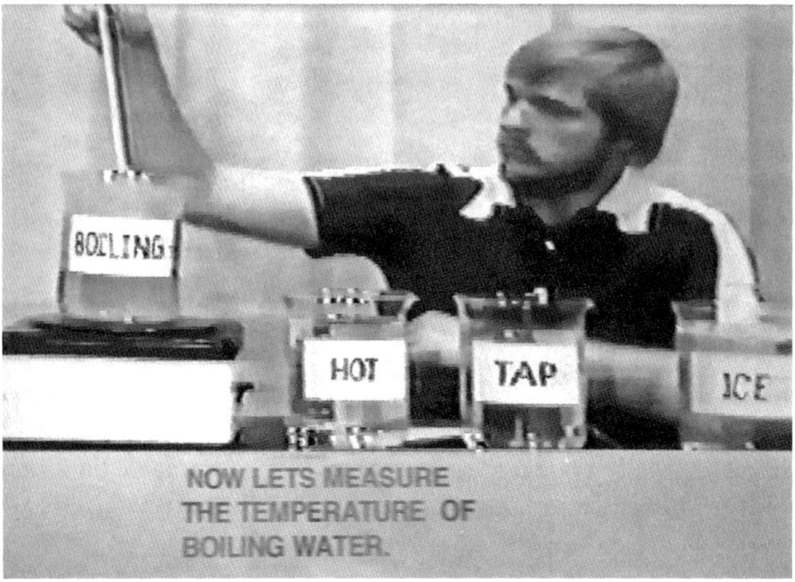


In 1974, there were not many captioned physics films. Author Harry G. Lang, a physics teacher at this time, developed a library of videotapes for his classes at NTID. Captions were printed on large index cards and inserted by hand. This particular film shows the author measuring the temperatures of boiling water, water at room temperature, and ice water. Image courtesy of Harry G. Lang.

other networks in offering captioned programs that could be accessed through the new TeleCaption decoders. It took years to have a menu of TV shows from which we could choose. I remember writing to one TV channel, asking why they broadcast foreign films with dubbed voices in English during prime time, but the same foreign films were shown with subtitles in the early morning hours. This was my first experience with activism. I was hungry to watch more films with subtitles. Unfortunately, the network explained in a letter that hearing viewers found the subtitles "distracting" and had complained about them. I was not at all surprised when I learned while working on this book how common that frustration was among other deaf people.

We also looked forward to reruns of TV shows, only to learn that the captions were often dropped. It seemed that the networks were compressing the films slightly for the reruns—enough to allow for more commercials—and this caused the spoken word and the captions to be out of sync. So the decision was made in some of these cases to just delete the captions.

While watching other shows, the captions would sometimes suddenly completely stop. Other times they were garbled. This was frustrating but, again, better than having no captions at all.

Even after decades of efforts by the federally funded program, the general public remained quite unaware of the importance of captioning for deaf viewers. When I. King Jordan became the first deaf president of Gallaudet University in 1988, he joined the efforts to educate the public. In the nearly three decades since John Gough described the potential of captioning, the technology had only inched forward, but the message remained the same. Jordan explained, "There's the notion that TV is a visual medium. Try watching TV some time with the sound off. . . . I promise you, you will soon learn that it's not very visual. It's really a sound-based medium with pretty pictures to make it more effective. Without the pictures, it can succeed. Without the sound, it can't."[2]

These words came from President Jordan twenty-three years after I had graduated from WPSD. And now, with degrees in both physics and electrical engineering (RIT in 1974), I was still without comfortable access on a level comparable to my hearing friends when watching television.

As with the costs of telephone access, we deaf people initially had to pay extra for special equipment to see the closed captions on television. I paid more than $275 for my first "decoder"—just to be able to see captions on very few programs. After I earned my doctoral degree in education from the University of Rochester in 1979, I was often invited to give presentations or to offer workshops on teaching. When traveling, I often brought with me the bulky decoder, which took up about half of the space in one suitcase. But at least I could watch captions on the televisions in hotel rooms. I also had to bring a set of tools to connect the cables myself. In many hotels, I attempted to request a room that had an accessible television, but it was not until after 1993 that hotels began to purchase new TVs with the built-in captioning decoder "chips" that were made possible by the Decoder Circuitry Act of 1990 and required by the ADA that same year.

In the 2000s, I weathered the days when theaters had Rear Window Captioning systems with small screens mounted on flexible arms that fit into the cup holders on the seats. Much of the movie time was spent adjusting the equipment to see the captions clearly. Eating popcorn or holding my wife's hand during a movie was out of the question. Keeping the captions visible on that small screen was a full-time job.

The same applied for the special glasses we would wear to read captions in some theaters. They are still being used as this book goes to press. Much of the dialogue would be missed as we attempted to locate the captions again after briefly looking away from the screen. More than a few times I left my seat in the middle of a film to inform the manager that the captioning glasses were not working. At times, the staff forgot to turn the captioning system on. Other times, the glasses were malfunctioning. I still have an email I sent to a friend: "Sadly, *Star Wars* yesterday was disappointing for me. The captioning glasses only worked on one side and the captions were weak. The lenses were very smudged. I told the manager to have them checked. He offered a refund but I declined to take it. I'll watch it on cable TV when it comes out months later."

Some hearing people may go for years never thinking about the importance of captioning and the impact that this technological breakthrough has had on the lives of deaf and hard of hearing people. But stopping for a drink in a noisy bar may help them quickly realize the benefits of such access as they watch a baseball game where the bartender was smart enough to turn on the captions for everyone.

Since the earliest days when silent films were shown in theaters, other benefits of subtitles and captions have been recognized. My best friend Bob Panara once told me about his parents, who had emigrated from Italy in the early twentieth century, and how they would go see the silents to help them improve their English. It seems a common practice today, too, with those interested in developing English skills. More recently, another friend of mine, Peggy Cagle, a hearing teaching specialist in Rochester, wrote to me, "I know I LOVE captioning on all the time. . . . between my ADD and tendency to multitask, I always have to check out what I just missed."[3]

Phil Bravin, a deaf former president of the NCI, humorously pointed to a downside of captioning: "There are some people that are couch potatoes now," he signed. "They are watching more television than ever."[4]

But captioning is definitely much more than an access issue. It is more than an assistive device in the classroom to aid in learning or in the theater for leisure time entertainment. Captioning has helped to assimilate us into American society by learning about how others live and function. Years ago, I went to see *The Theory of Everything* in a theater. I had met Stephen Hawking during his visit to America in 1984 when he interacted with other scientists with disabilities. I had read his books and followed his theories. Yet, I had no idea how similar his personal experiences in adjusting to a motor neuron disease were to my own experiences as a boy who survived a coma and began a new life totally deaf. I was deeply moved by the film.

The captioning program has opened up avenues not only to recreation but also to such learning experiences. As the pioneers in the early 1960s had hoped, captions now convey, as Mac Norwood, the first deaf chief of the Media Services and Captioned Films Branch, US Office of Education, said, "developments in language usage, new terms, new idiomatic expressions, and social changes to deaf persons in such a way that their vicarious experiences are expanded into the real world."[5]

Today, deaf audiences have wider access to a variety of film entertainment. Through the use of closed captioning technology, television is largely accessible to deaf people. Although access to movies in theaters continues to be limited, deaf people can now view most major films with captions by purchasing or renting them a few months after their general release to the hearing public. Streaming videos, in particular, has become very common, with most of the major streaming services now including captions in their offerings.

In July 2019, I was traveling home to Rochester on a Delta Airlines flight from Athens, Greece, during which, on the screen mounted at the back of the seat in front of me, I watched three captioned films. I was pleased about the access on that long flight. I thought about Delta Airlines for a moment. I had once written about a deaf entomologist from Cleveland, a former student at my alma mater, WPSD, whose pioneering flights in crop-dusting airplanes in the spring of 1921 contributed substantially to the evolution of aerial applications aircraft, the basic aircraft in the Delta Air Service, later to become Delta Airlines Company. The airline had come a long way, I thought.

Only a few months later, however, the Deaf actor and captioning advocate Marlee Matlin tweeted, "Sad to see that my preferred airline, @Delta flight 1998 has provisions for various languages and audio description for in-flight entertainment but no closed captions for deaf and hard of hearing flyers."[6] Her post quickly received a million views and 48,000 likes. In the three decades since the ADA was passed, "missing closed captions—or incorrect, usually automatic captions known in the community as 'craptions'—have continued to be problematic for deaf and hard of hearing audiences in many venues, both in the air and on the ground."[7] As Sara Nović explained in an article titled "The Frustrating Fight for Closed Captions on In-Flight Entertainment," the ADA doesn't regulate air travel. Earlier, in 2015, Deaf model and actor Nyle DiMarco also sought access to in-flight entertainment. He tweeted, "Myself and the 466 million other deaf people in the world aren't the only ones who benefit from captions . . . [they improve] viewing experiences for English language learners, autistic people and people with ADHD and learning disabilities." Matlin had worked hard in the early 1990s to help pass legislation for closed captioning by speaking up on talk shows and news programs. Now, in 2019, she pledged to continue to raise awareness about inequity in air travel. "Today, platforms like Twitter and Instagram serve the same function. Until airlines come up to speed when it comes to providing access for travelers who are deaf and hard of hearing, I will not stay silent."[8]

But watching the three films on my own flight from Athens, I was also reminded of the discussions in this book about the lack of accessibility to open captioned films in movie theaters that continues to this day, six decades after the pioneers had begun working on access through captioning. One of the films I watched, *The Aftermath*, focused on postwar Germany in 1946, where a British colonel in Hamburg is charged with rebuilding the shattered city. Fiction or not, the film has educational value. There is much to learn about how families were destroyed, for example, not just cities. The film provided a different perspective that many classes in World War II history do not cover.

Although it is true that the captioning pioneers had led the way in making educational films accessible through teacher-led experiences, the opportunities to learn through *vicarious* experiences, especially fictional films in theaters, remain somewhat restricted for deaf people of all ages, which is more than they should be after six decades. The menu on the screen I was watching, for example, had many films listed that were not accessible through captions. Granted, there are first-run movies and many second- and third-run movies fully captioned on television, and many more are available for an extra cost on premium channels such as HBO, Cinemax, Starz, and Showtime, as well as Netflix, Amazon, and Hulu streaming services.

As resources for open and closed captioned films continue to be explored in the future, the technologies developed for K–12 students, especially streaming, bilingual accessibility, and description, will certainly be available for others to expand for different audiences outside the educational arena.

The anecdotes I have shared in this introduction represent a small portion of a trove of experiences that deaf and hard of hearing people have had with captioning. I hope that the following pages will develop a much greater appreciation of the hard work and dedication of hundreds of pioneers who "turned on the words" for us.

# 1
# The Golden Era

*Perhaps, in time, an invention will be perfected that will throw the words spoken directly under the screen as well as being spoken at the same time.*
—Emil Ladner, "Silent Talkies," 1931

Throughout the nineteenth and early twentieth centuries, teachers dreamed of using innovative media that could enrich the education of deaf children. For decades they had experimented with combining pictures and words in their classroom instruction. In 1870, for example, James H. Logan, a deaf teacher at the Illinois Institution for the Education of the Deaf and Dumb, noted the usefulness of adjunct aids to improve learning. "Pictures," he wrote, "besides the pleasure they give, act as definers of the text, and convey far more correct ideas than could be gained from the words alone."[1] Long before motion picture films arrived, he recommended the use of an "illustrative apparatus," such as the stereoscope, stereopticon, or magic lantern slides for enhancing learning and entertainment through the combined use of pictures and text. "There is no good reason why every institution should not be provided with a good library and some illustrative apparatus," he wrote from Jacksonville. "All the pupils, even the youngest, may be made to comprehend, but in different degrees, such things as they serve to illustrate. . . . The expense should be considered no objection at all. . . . The system of deaf-mute education must keep pace with that for the hearing and speaking . . . "[2]

## THE FILM INDUSTRY BEGINS

It is fitting that the history of the entertainment and educational film experiences of deaf people includes the work of a deaf inventor, Thomas Alva Edison, who announced at his laboratory in October 1888 his plan to invent "an instrument which does for the Eye what the phonograph does for the Ear."[3] Edison's intention was to record and reproduce things in motion. The motion picture films he first developed were not meant to be primarily educational.

Edison, who has been described as America's greatest inventor, fostered a vision at his West Orange, New Jersey, laboratory to develop the motion picture to help "liberate the creative spirit by crystallizing and applying a teamwork approach to research and development."[4] The technology he developed, however, ushered in a revolution that provided a roller coaster of emotion for deaf people over the decades to follow.

Born on February 11, 1847, in Milan, Ohio, Edison grew up in Port Huron, Michigan. He only attended school for a few months and was subsequently tutored by his mother. Edison credited his deafness with helping him to concentrate on his experiments.[5] He went on to accumulate 2,332 patents, leading Joseph Henry, president of the National Academy of Sciences, to describe him as "the most ingenious inventor in this country . . . or in any other."[6] His first motion picture apparatus, developed with the assistance of his employee, the photographer "William K. L. Dickson, featured a series of small photographs mounted on a cylinder not unlike that of his [early] phonograph. The images were viewed through a microscope attached to the rotation cylinder."[7] Edison focused on the electromechanical design while Dickson worked on the photographic and optical development and deserves much of the credit for the invention.[8] In particular, Dickson was given the task of working on the process of moving film through the camera and propelling the image through a machine to view the picture—the Kinetoscope—the forerunner of the motion picture film projector.[9]

The film was obtained and codeveloped through collaboration with American inventor and manufacturer George Eastman of the Eastman Kodak Company in Rochester, New York. After more work, patent applications were made in 1891 for a motion picture camera called a Kinetograph.

During the winter of 1892, Edison's team constructed the world's first motion picture studio. Nicknamed "Black Maria," it was the location for the filming of Fred Ott in *The Sneeze* (1894), a boxing match between James Corbett and Peter Courtney, and *Serpentine Dance* (1894) starring Annabelle Whitford. Edison's team also made documentary chronicles of real life, such as *New Brooklyn to New York via the Brooklyn Bridge* (1899).[10]

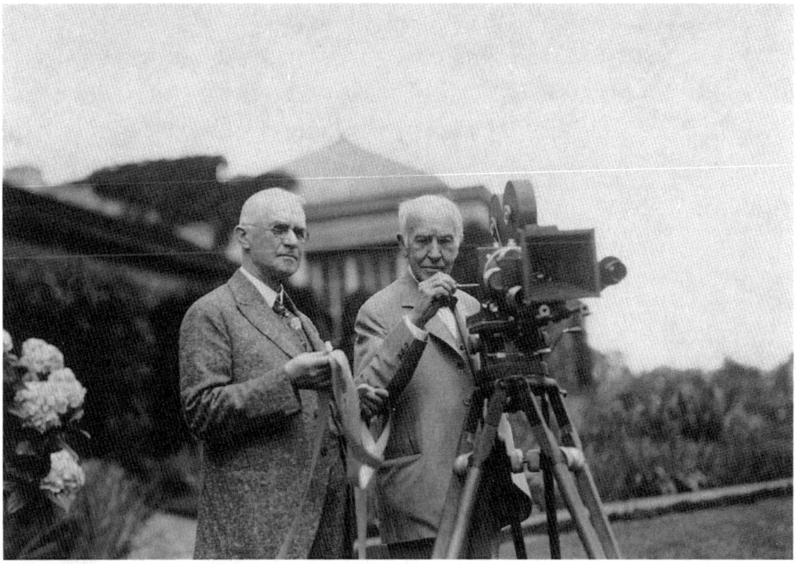

George Eastman (left) and Thomas Edison with a motion picture camera at Eastman's house in 1928 in Rochester, New York, where a demonstration of the new Kodacolor film was being held. Image courtesy of the Library of Congress.

On April 14, 1894, the world's first commercial demonstration of Edison's Kinetoscope took place at an arcade on Broadway in New York City. Within months, Kinetoscope or "peephole" parlors, also known as "nickelodeons," opened in cities throughout the United States and Europe, including San Francisco, Atlantic City, and London.

Then, in 1895, two Frenchmen took advantage of Edison's failure to patent his peephole machine in Europe. Louis and Auguste Lumière began to manufacture the Cinematographe, a more portable camera that projected moving pictures onto a screen. Their initial productions were realistic depictions of scenes from daily life and proved to be popular. As the collaboration between Edison and Eastman moved ahead, Étienne-Jules Marey's work with roll film in France further enhanced the development of motion picture entertainment. The Lumière brothers closed their business in 1905, but other competitors such as the Biograph Company and Georges Méliès's Star Film Company entered the field.[11]

Meanwhile, the deaf artist and photographer Theophilus Hope d'Estrella was creatively pursuing the use of still pictures. Known affectionately by students at the California School for the Deaf as the "Magic Lantern Man," d'Estrella used lantern slides, glass-plate positive transparencies that were either black-and-white or hand-colored with transparent oils, and projected the images onto a screen using the stereopticon. The first Magic Lantern Exhibition held at the California School for the Deaf occurred in February 1888, and Berkeley citizens were invited to attend. "In the following weeks," wrote d'Estrella's biographer, Mildred Albronda, "the children applauded the blood-and-thunder images of great battles of the American Revolution and of the Civil War on land and sea. . . . Some of the slides produced the effect of motion which was accomplished by a special arrangement in the machine."[12] Before long, d'Estrella was presenting thirty to forty lantern-slide lectures per year to the deaf children.

As early as 1903, however, fascination with the lantern-slide magic was replaced by interest in the new moving pictures, and by 1910 the California School for the Deaf purchased motion picture equipment and was showing two or three movies a month. D'Estrella himself envisioned that the motion picture film would bring widespread changes in people's use of leisure and in their perceptions about leisure time. Other schools serving deaf children soon followed. In Arkansas, the school's baseball team raised funds to purchase a moving picture projector in 1912 by playing a game with the employees of the State Hospital for Nervous Diseases. Superintendent Isaac B. Gardner saw "wonderful educational value" in motion picture films for deaf pupils.[13]

The Edison Manufacturing Company sent crews around the world to capture unusual sights and events and began to produce more complexly plotted films. In December 1903, the company released Edwin S. Porter's popular hit, *The Great Train Robbery*, which had location shots filmed near Edison's laboratory in New Jersey. Many subsequent entertainment films were produced at Edison's studios in Manhattan (1901–1907) and the Bronx (after 1907). Around the same time, other movie studios began competing with Edison. Eventually these studios moved many of their operations to Hollywood, California, and soon came to dominate the industry.

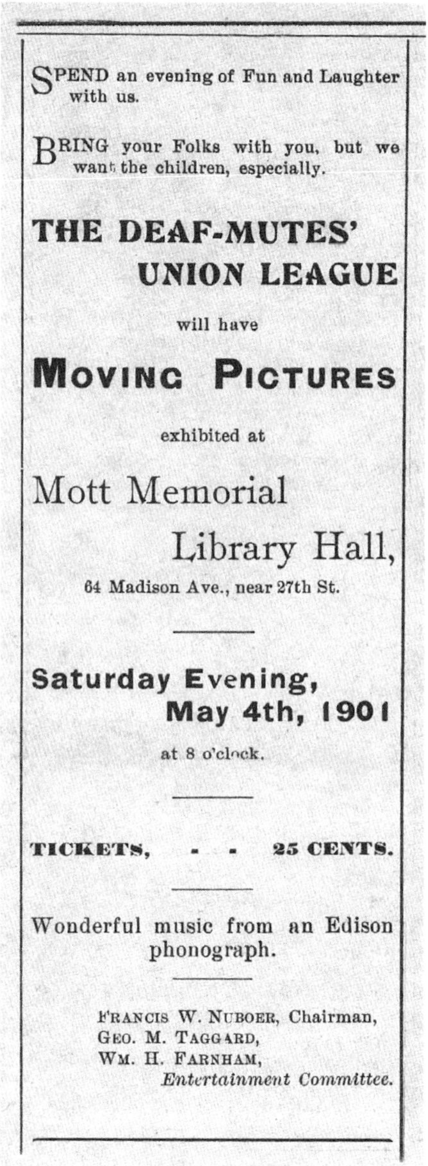


By the turn of the century, deaf people were enjoying the "moving pictures" at social gatherings. Some with partial hearing also appreciated the accompanying music played on Edison phonographs. This advertisement was for The Deaf-Mutes' Union League Club gathering on May 4, 1901, in New York City. Image courtesy of the Gallaudet University Archives.

In addition to Edison's studio making close to 1,200 films, in 1908 he started the Motion Picture Patents Company, a conglomerate of nine major film studios. Four years later, he introduced his "Home Producing Kinetoscope," one of the first home projection systems.

During the first two decades of the twentieth century, silent films became a popular attraction for both deaf and hearing people. With their visually effective quality and the use of body language and facial expressions, the silents presented an enjoyable medium for entertainment in the theaters. In addition, deaf actors were able to find employment, and schools serving deaf children saw the growing potential of motion picture film for education. In 1915, for example, the New Jersey School for the Deaf showed weekly films on topics such as nature study and bookmaking, and at least a dozen other schools owned projectors.[14] "Our motion picture program becomes more extensive and important from year to year," reported the Indiana School for the Deaf in 1929. "We are finding new ways to coordinate the motion pictures with our school program."[15] This school secured newsreels and longer plays from local exchanges, and educational films from Indiana University. The films were well received by the children and became part of the curriculum of the school day, whereas longer entertainment films were shown outside of school hours. The school reported that the educational films became "the basis of much of our composition, Geography, History and Current Topics, and the pupils followed up by requesting such items as wall maps in order to look up the locations of countries mentioned in the films."[16]

In his book *Hollywood Speaks: Deafness and the Film Entertainment Industry*, John "Stan" S. Schuchman described how the silent film era "inadvertently" included deaf people who participated as equal members of an audience, as beneficiaries at schools, and as actors on the screen.[17] He referred to this period in the cultural history of the American Deaf community as the "Golden Era." Previously, sign language could only be expressed in static form. Now, movies were becoming a cultural product for deaf people. According to Schuchman, deaf audiences especially appreciated the silent films for the action and the expert use of facial and body expressions by many actors and actresses. Silent films depended little on the use of words in telling a story and provided a new tool for those who used sign language. "This explains the success of Charlie Chaplin," Schuchman wrote, "who conveyed whole sentences with the twitch of an eyebrow."[18] Chaplin's film subtitles were few and short.

## PRESERVATION OF SIGN LANGUAGE

The age of silent films happened at the same time as influential people were campaigning against sign language in the United States,[19] and films also were used as a form of social resistance during this period in history. Since the late nineteenth century, there were efforts to ban the use of sign language in American schools. Advocates of oralism emphasized the use of speech, speechreading, and residual hearing for educating deaf children, and they demeaned communication through sign language. Deaf people unable to teach speech were losing their positions in the schools, and the Deaf community, perceiving this as an attack on their linguistic rights, fought bitterly against the suppression of their visual language. They sought to preserve sign language through the use of motion picture technology. Between 1913 and 1920, the NAD collected funds to produce a series of 35 mm films on which were recorded poetry, lectures, memories, and stories in American Sign Language (ASL) as signed by deaf and hearing masters of the language. The films were later converted to the more durable and less flammable 16 mm safety film and deposited with the Library of Congress and Gallaudet College.[20]

George William Veditz was a leading activist in the American Deaf community who battled the suppression of sign language by Alexander Graham Bell and other proponents of oralism. In the film *Preservation of the Sign Language* (1913), Veditz argued that "As long as we have deaf people on earth, we will have signs. And as long as we have our films, we can preserve signs in their old purity. It is my hope that we will all love and guard our beautiful sign language as the noblest gift God has given to deaf people."[21]

By the 1930s, some members of the Deaf community were also taking advantage of film and cameras designed for the amateur home film market. This was, in effect, the beginning of the home film market for deaf people. Social events and athletic contests were popular topics to record. One deaf man from Chicago, Charles Krauel, filmed visits to schools serving deaf children, conversations with deaf adults in various locales, and meetings of the National Fraternal Society for the Deaf, the national self-help insurance organization for deaf people in the United States.

Without the need for audio, deaf actors and some hearing actors such as Charlie Chaplin did not even mouth the words in the silents. One silent film, *His Busy Hour* (1926), directed by James Spearing and produced by Bertha Heustis, even featured an all-deaf cast.[22] Deaf people slowly became accustomed to equal access at movie theaters and gained awareness of the participation of deaf actors such as Granville Redmond. Another was Emerson Romero (stage name Tommy Albert), whose cousin was the famous movie actor Cesar Romero. Altogether, there were five deaf actors appearing in silent films consistently.

## THE TALKIES ARRIVE

In the mid-to-late 1920s, advances in the commercialization of sound cinema included synchronized dialogue, and these films were known as "talking pictures" or "talkies." These were mostly "shorts" (usually defined as an original motion picture film with a running time of forty minutes or less). *The Jazz Singer*, released in October 1927, was the first feature "talkie" film.

The Deaf community, heretofore entertained on an equal basis with hearing people, was now unable to completely follow the action in the talkies. The reaction was swift. "The totally deaf," reported the *Deaf-Mutes' Journal* in January 1929,

> and also the hard of hearing, are beginning to be somewhat shy on patronage at the 'movie' theatres where sound effects are an addition to the motions on the screen. It is quite natural, if not reasonable, that the deaf should shun the 'talkies,' which eliminate the enlightening prints of explanation that alternate with the pictured scenes.[23]

The writer in the *Deaf-Mutes' Journal* also claimed that the talkies were reducing the action to allow for more music or speaking. The "sound pictures" were largely relying on orchestral music, and usually in the interspersed dialogue, the characters were considered by some as "'stiff' and immobile."[24]

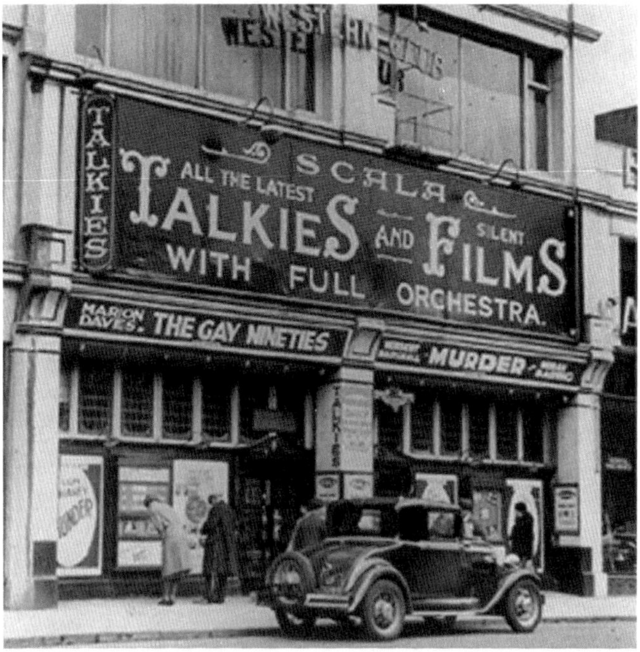


The advent of "talking picture" films, which include dialogue, music, and other forms of sound, transformed the American film industry and mass entertainment. The first feature film originally presented as a talkie was *The Jazz Singer*, which premiered on October 6, 1927. Image courtesy of Brighton Museum, UK.

Even deaf and hard of hearing people who were considered good lipreaders were unable to understand most of the speech of the "dialogue characters." An actor who spoke quickly or whose lips were obstructed (such as having a hand covering the mouth, a mustache, the speaker turning his back, or off-camera voicing) further decreased understanding. Members of the Speech Reading Club of Philadelphia circulated a petition to clubs for hard of hearing people; the signed petitions requesting captions were sent to motion picture producers. Principal Frank R. Wheeler at the American School for the Deaf (ASD) also expressed deep concern about the complete displacement of the silent films, seeing a "great hardship" brought to deaf people. But he felt that a "minority protest . . . will not do much good."[25] "It is futile to protest against anything so universal as the sound pictures," declared Mrs. Harriet Montague, staff writer for the *Volta Review*, a magazine for people who communicated primarily through speech and speechreading. She saw such a protest as comparable to an organized objection to the radio, which was also enormously popular.[26] Yet, if it meant even a few silent movies could be saved, she would sign anything.[27]

According to an estimate issued by the Volta Bureau in 1929, there were about 10 million hard of hearing adults in the United States and 3 million children of school age. The number of adult deaf individuals was placed at 44,885, and the number of children at 17,192, bringing the total of those with deafness to more than 13 million.[28]

Sound quickly consumed the film industry, and as 1929 drew to an end, 8,741 movie houses were equipped for sound. In addition, snatches of quickly recorded dialogue were urgently inserted into productions that were already shot as silent.[29]

The 1920s also witnessed some of the earliest experiments in television, as well as the advent of radio, which exacerbated the frustration deaf people already experienced in having no access to the common household telephone. It seemed to some that the talkies were another technological advance that was primarily a boon for hearing people but another nemesis faced by the Deaf community. Presciently, as the editor of the *Deaf-Mutes' Journal* summarized, "It may be that, in the not far distant future, the radio and the wonders of television will combine to bring the theatre into the homes of all who can afford to install the necessary receiving outfit."[30]

## THE CALAMITY OF SILENCE

All over the country, activism grew against the talkies as objections continued to be raised by deaf and hard of hearing people. Letters were sent to the press, and protests were registered with the producers of sound pictures. Dissatisfaction with the new films was expressed in meetings held in many cities. "We are not asking that the 'talkies' be eliminated," one writer explained, "but we are making the request that enough captions be included in all the films to enable us to follow the general plot of the play. We do not believe that the insertion of captions will make the play less interesting for our hearing friends, and they will serve as the link to hold the gradually wavering interest of deafened patrons of the hitherto silent drama."[31]

Marcus L. Kenner, vice president of the NAD, said, "The advent of 'talkies' obviously makes screen interpretation difficult for us, with the result that almost our sole prop of enjoyment is ruthlessly removed. It is like taking away a blind man's stick [cane] and offering him no support in its place."[32]

At first, the general public did not realize how deaf people were losing out with less action and the removal of captions. Blind people were benefiting, however, because they could hear the soundtrack on the films.

One argument against keeping the silents was that the general public was tired of them, perhaps because of the sameness of the film fare, or perhaps it was that people were being weaned away by radio. No matter what the cause, box office figures proved that in order to make a big motion picture theater pay profits, it was necessary to reinforce the screen with an expensive stage show. The talkies brought the hearing public back to the theaters in droves.

Various adjustments to the talkies were attempted by different groups. In 1930, three Chicago theaters installed electrical hearing aid equipment that would enable the 300,000 hard of hearing people in the Chicago area to gain better access. Of course, this was little help to those who lacked residual hearing.

Some deaf people watched foreign films with subtitles. Others continued to try to lipread or guess at the context of the films. It was no surprise, then, that some people in the Deaf community consequently turned to lectures, skits, plays, and dances held by organizations serving deaf people, and to viewings of old silent movies whenever possible. Unfortunately, these films eventually deteriorated and projector technology advanced beyond the stage where the apparatus could be used with the older films.

Despite these attempts being such poor substitutes for the silent films that were previously enjoyed, deaf people understood that the talkies represented an improvement in film technology. "The deaf do not stand in the way of progress nor bewail any advance that entertains those who can hear," the editor of the *Deaf-Mutes' Journal* explained, yet he continued to express concern about the dropping of subtitles.[33] He was also critical of the quality of the acting: " . . . it is difficult to understand how the hearing world can get actors who not only 'screen well,' but also talk well. It would seem that the elocutionary effect, or the screen excellence, must suffer; because great actors, who sway multitudes by the sound of their voices, combined with incomparable stage presence, are very rare."[34]

Thomas Edison's favorite silent film was *The Birth of a Nation*, and his favorite silent film stars were Mary Pickford and Clara Bow. He also thought that talkies "spoiled everything" for him. "There isn't any good acting on the screen. They concentrate on the voice now and have forgotten how to act. I can sense it more than you because I am deaf."[35]

And Charlie Chaplin loathed the talkies:

> They are spoiling the oldest art in the world—the art of pantomime. They are ruining the great beauty of silence. They are defeating the meaning of the screen . . . the vast popularity of the whole—the appeal of beauty. . . . It is beauty that matters in pictures—nothing else. The screen is pictorial. Pictures . . . I am not using the talkies in my new picture. I am never going to use them. For me, it would be fatal.[36]

Schools serving deaf children continued to use older silent films both for educational and entertainment purposes whenever possible. As Shuchman writes, "the advent of talkies impeded the flow of new information about the world to deaf students. Not only was the potential for English language development of deaf students through the use of silent films lost, but these students were also further isolated from the culture of a hearing American society."[37]

The first arc-projected 16 mm educational film came out around 1930. Although this technology, the lantern slides, and other innovations brought some supplementary visual aids to instruction, it was not until Public Law (PL) 85-905 in 1958 establishing CFD that a government-sponsored breakthrough began an era of access to visual representation of the spoken word. Although demonstrating occasional ingenuity, the level of media utilization in schools serving deaf children was heretofore sparse. The change beginning in the early 1960s was amazing.

CFD

Captioned Films for the Deaf

Captioned Films for the Deaf (CFD) logo. Image courtesy of Described and Captioned Media Program.

Emil S. Ladner, a Gallaudet College preparatory class student from Cupertino, California, was not yet nineteen years old when he won a prize in the *Atlantic Monthly* essay section of the National High School Awards Contest. His essay in 1931, written while he was a pupil at the California School for the Deaf, was titled "Silent Talkies" and expressed well the hopes and dreams of deaf people faced with the effects of this technological impact. Frustrated by the "calamity" of the talkies depriving deaf people of entertainment and education, young Ladner wrote, "Perhaps, in time, an invention will be perfected that will throw the words spoken directly under the screen as well as being spoken at the same time."[38]

# 2

## The Pioneers

*They also said it could not be done. . . . I decided to do it myself just to show them it could be done.*

—Emerson Romero, deaf filmmaker, quoted in Dorene Romero, "Emerson Romero: That's My Pop"

The early years of the electric telegraph and silent films represent brief times in our history when deaf and hard of hearing citizens shared comparatively equal access with hearing people. Technological advances such as the telephone introduced dramatic changes in this type of access. It took nearly a century for deaf people to gain regular access to this electrical form of telecommunication.[1]

Similarly, after the advent of the talkies in the late 1920s, it took several decades before captioned films, as we know them today, were enjoyed by deaf people on a regular basis. The seminal changes in the technology to caption films occurred as a result of the dedicated efforts of both deaf and hearing people who saw the importance of equal access to both entertainment and educational films. The dire need of such access to a visual representation of spoken messages led these pioneers to apply their inventive talents and explore alternatives.

Emil S. Ladner's dream of captioned films was, of course, one that many deaf people shared. Sylvia Sanders, a graduate student at Gallaudet College, expressed the dream again in 1946. She wrote presciently about how a special motion picture library might be established from which different schools for deaf children could draw. "Such a library would lend free and rental films . . . as books are lent."[2] Sanders also described how teachers could exchange copies of materials they composed for use with the films. Importantly, she wrote, "If such a center were well organized and established, it might be able to exert an influence on commercial and non-commercial film companies. It could urge that pictures be retitled and captioned to contain new words and build vocabulary without being beyond the comprehension of the children."[3]

## EMERSON ROMERO'S CAPTIONING EXPERIMENTS

The first effort to caption films for deaf viewers was a revival of the technique of subtitling used before the talkies took hold. In New York, Emerson Romero, one of several deaf actors who were unemployed with the advent of talkies, became involved in establishing a professional deaf repertory company, the Theatre Guild of the Deaf.

Romero first attempted to convince the producers and distributors of 16 mm films to insert subtitles. "They did not want to do that. . . . They said the market was not big enough. They were in business to make profits. . . . They also said it could not be done. . . . I decided to do it myself just to show them it could be done."[4]

At the time of his captioning experiment in 1947, Romero was working for an aircraft manufacturing company. His brother helped him obtain a few old films to try to adapt them for deaf viewers. Frustrated with the inaccessibility of the talkies, he began to splice in subtitles using dialogue cards, and, on a shoestring budget, he was purchasing sound films and renting the resulting films out to deaf organizations and churches.[5] The quality of the films was poor. Producers were also worried that they might be used as bootleg copies.

Romero established the National Film Library to enable deaf and hard of hearing people to derive the full benefits from films. He first previewed 16 mm films and deliberately chose those that had enough action to be entertaining and included minimal dialogue. Scripts were then obtained from the producers, and "unnecessary dialogue" was eliminated to reduce the number of subtitles. The printing department then printed the subtitles on cards using a thirty-point boldface type. At times, the words were edited for young deaf children but sometimes also for deaf adults who might not be familiar with certain dialogue, such as lines that might be uttered in Western films (i.e., "I ain't gonna tool this six-up into town" was changed to "I won't drive this six-horse coach into town"). Some feature films incorporated as many as 500 subtitles, and care was taken to allow at least one second for every two words. The subtitles were then inserted as the film was spliced. The first film selected for Romero's National Film Library was *Rebellion*, which included about 467 subtitles to ensure that the film "actually became alive" for deaf viewers.[6]

The trade-off for access at this time was a reduction of quality. The process of superimposing subtitles was not yet developed well enough, and the "stop and read" procedure of blanking out portions of the frames while shooting and then reshooting with the prepared text on the unexposed areas was complicated, costly, and seldom used.[7]

During the first three years, the National Film Library received sixty-eight bookings of its three feature films from schools and Deaf clubs around the country. One exhibitor wrote: "The film was very good and all those who saw it enjoyed it immensely. The subtitles helped them to understand the entire picture."[8] Romero also took on the responsibility to personally find projectors for any Deaf clubs having trouble obtaining them.

In 1957, Romero announced the affiliation with Nu-Art Films, Inc., one of the leading distributors of sound and silent films. This was the eleventh year of the National Film Library.

Romero's method, however, was unsatisfactory. The subtitles not only interrupted the flow of action of the film and the dialogue of the actors, but the overall process also lengthened the movie. Yet, despite

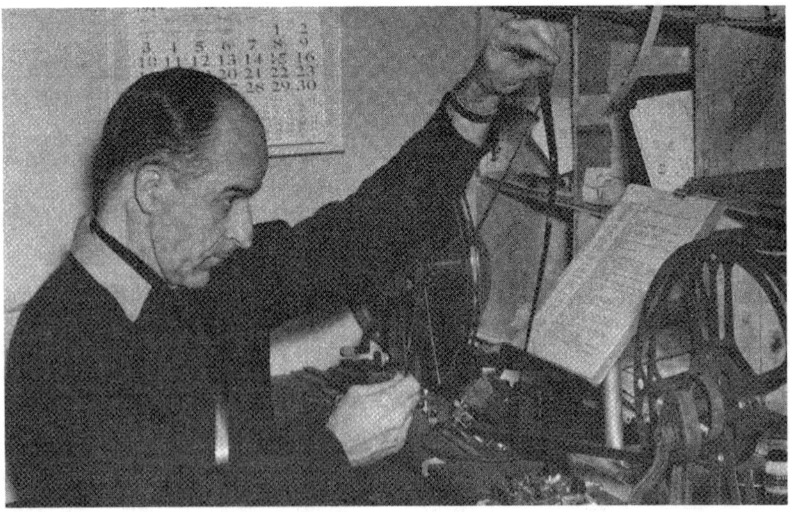

Emerson Romero, pioneer in captioning films for deaf people, is shown checking 16 mm film frames against a list of captions while working in his home movie shop in 1949. Image courtesy of Wallace Memorial Library at Rochester Institute of Technology.

the reluctance of producers to make better prints available because of the danger of piracy and the heavy work needed to procure, prepare, and circulate the films, as well as the lack of adequate funds, Romero was able to provide access for deaf Americans to a few of the movies being produced at this time. It is amazing that he accomplished as much as he did. And, as Bill Stark later wrote in an article about Romero and other captioning pioneers, "We should never forget that the dream of captioned movies for all of us originated in the mind of a deaf man."[9]

## ERNEST MARSHALL'S CAPTIONED ASL FILMS

The independent deaf filmmaker Ernest W. Marshall of New York City was recognized by Deaf community theater actors for his potential as both an actor and a sign language master.[10] In the 1950s, Marshall also experimented with adding captions to his ASL films, making them accessible to hearing and deaf audiences, both signing and oral, and giving them a choice of language. His method was tedious and time-consuming. He would rewind the exposed film and then expose the film again to black title cards with white, hand-painted English subtitles precisely placed. He also experimented with the "bouncing ball" technique to allow viewers to follow the song with the subtitled words in the short film *I Shall Know Thereafter.*

Between 1956 and 1959, as educators of deaf students were working toward the establishment of an organization to caption and distribute films, Marshall was pioneering in directing and adding dialogue subtitles to such films as *The Confession*, *The Face on the Barroom Floor*, and *The Debt*. In 1960, Marshall and Romero collaborated on the film *A Cake of Soap*, with Ernest filming and Emerson directing.[11] As George Potanovic summarized:

> [Marshall's] true reward was realized as he saw the expressions of excitement on the faces of deaf audiences when they watched his films. He was not satisfied to simply ship his films to Deaf clubs and wait for their return. He wanted to be there to experience the showing of his films.[12]

These deaf filmmakers satisfied an unmet need for entertainment in the Deaf community.

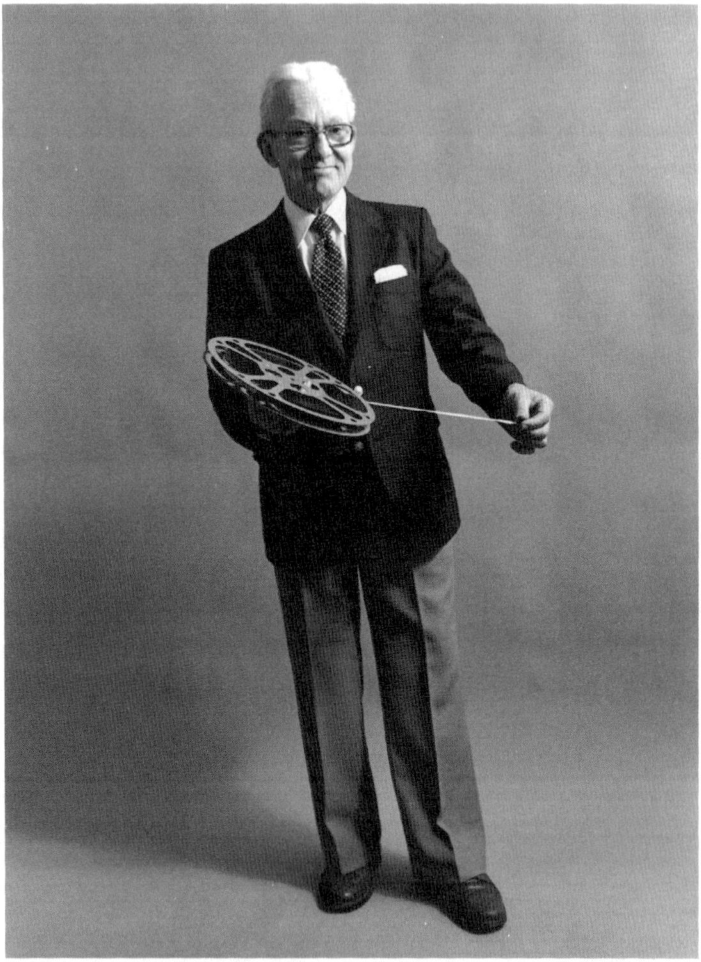

Ernest W. Marshall often used deaf actors with stage experience in creating sign language films. While the earlier NAD films incorporated masters of sign language, Ernest Marshall's films featured deaf actors who used signs. Image courtesy of George Potanovic.

## ROSS HAMILTON'S "VISUAL NARRATIVE"

In the late 1940s, several educators began to envision different approaches to captioning films. Existing technology remained limiting. Ross Hamilton, a hearing administrative assistant to Clarence D. O'Connor, director of the Lexington School for the Deaf in New York City, was serving an internship as one of the original doctoral candidates at Teachers College, Columbia University. One of his assignments was to experiment with educational films for possible use with deaf students. He was keenly aware of the language development issues of deaf children at this time, and his interest in visual aids brought a number of staff at the Lexington School to assist him with the project. In a letter dated May 21, 1947, to John K. Norton, his professor at Columbia, Hamilton described his experiment: "The third project can be described as adaptation of sound movies for use in schools for the deaf. The need for this grows out of the fact that silent movies are becoming increasingly rare and of those produced, most are technical and out of our range of interests. The cost of producing top notch educational silents for the deaf is also 'high' as the market for the product is limited. Adaptation of what exists, although often not ideal is our way out."[13]

Hamilton explained that the project was still in the early developmental stage. "For the present, we are preparing a visual narrative on lantern slides. These are superimposed on the screen over the picture from the sound movie and incidentally, this involves a little technique to overcome the 'double exposure-over exposure' result which we first obtained on the screen."

He encountered many technical problems. "Timing and duration of the visual narrative is a delicate matter but is achieved with encouraging success. To reduce the strain on the lantern-slide operator, two of these machines are now being used with a home-made 'fade-in fade-out' device which results in a smoother visual narrative." Training a group of teachers to prepare a script was in itself a big chore. "They started with subtitles, emerged later with a translation of the sound story and now are doing a visual narrative entirely in terms of the presentation to deaf children."[14]

The method Hamilton was demonstrating was similar to that being used by the Englishman J. Arthur Rank, the millionaire flour miller who bought up the Odeon chain of cinemas. Rank's organization in England was controlling the production, distribution, and exhibition of films. His method involved projecting the captions from a second, separate projector onto a smaller screen in the lower left corner of the main screen. Hamilton's captions, however, were projected over the sound film picture. He examined subtitled foreign films, the work of

Emerson Romero, and other films. The advice he received from others was to "tell it in narrative, but present everything through the eye." Hamilton thus referred to his captioning technique as "visual narrative."[15]

He subsequently produced a visual narrative for *America the Beautiful*, a Warner Brothers color release. At the Convention of American Instructors of the Deaf (CAID) meeting at the Florida State School for the Deaf in 1947, he sparked much interest among teachers who saw the far-reaching potential of captioned motion picture films. These teachers were willing to ignore the cumbersome technique with two projectors and enthusiastically requested training through workshops.

"Ultimately," Hamilton wrote to Professor Norton, "we hope to produce a negative of a movie to which has been added a visual narrative."[16] The young man was a visionary planner, and his thoughts extended to the possibility of applying this process to entertainment films. But as Romero also learned, the costs for an individual, or even a school, were prohibitive.

## THE CAPTIONING ENTERPRISE: THE PROBLEMS OF START-UP

Intrigued by Hamilton's experiment in his school, O'Connor subsequently contacted Edmund Burke Boatner, superintendent of ASD in West Hartford, Connecticut. Boatner also gave much consideration to the popular reception by British deaf people of a captioned film produced especially for them by Rank in the spring of 1950. John Mills's film *Morning Departure* was projected normally while a second projectionist stood by with 350 lantern slides bearing captions. Every fifteen seconds the projectionist changed them, so that the action was flashed on the bottom of the screen "providing an almost continuous synopsis." Rank's experiment generated much enthusiasm among the 500 people invited to the film showing. The deaf moviegoers crowded the foyer afterward, chattering excitedly in sign language about the new breakthrough. One newspaper reported that it had been decided to prepare "slide scripts" of other films as soon as possible in order to show films to deaf people all over Britain. "For more than 250,000 Britons who stay away from cinemas because of their [deafness], this will open the door to a new entertainment."[17] There were also whispers of captioning

Hedy Lamarr's *Ecstasy,* in which she appeared nude, and that a "purified" version would be available for that purpose.

Another film captioned by Rank, *Crash of Silence*, was exhibited at the Lexington School, where the same audience that viewed Romero's *Rebellion* preferred the continuity maintained through the screen time as compared to the use of subtitles that separated the scenes.

Boatner long wanted to see such access through captioning. Three years earlier, in 1947, he took the ASD basketball team into town for dinner and the movie *The Son of Monte Cristo.* He recalled the unsatisfactory situation he observed that evening:

> As I watched the boy's reactions, I could see the looks of bafflement on their faces. In one scene, for example, a group of men were casually sitting around a table talking when suddenly they jumped up and started in at one another with their swords. Why? Our boys couldn't see any reason for such behavior; they hadn't heard the conversation. It was then that I made a resolution to see that understandable films were provided for the deaf.[18]

However, when Boatner visited the Lexington School and saw the short film captioned by Hamilton, he concluded that it was not a practical way to caption films. Whether two projectors were used with one screen or two screens, it was just not feasible. Rank felt the same way about his own efforts in England.

Boatner was nevertheless impressed by the worthiness of the attempts, regardless of the technical difficulty and the clumsiness of keeping the two projectors in synchronization. Yet he knew that something better was needed. Commendable as the efforts by Marshall, Romero, Rank, and Hamilton were, these pioneers had little chance of succeeding. The unwieldy methods of adding captions were compounded by the lack of adequate funds for acquisition, preparation, and distribution. "Obviously," Boatner wrote later, "there should be films with captions, but how were they to be made?"[19]

Importantly, although Boatner hoped to obtain educational films adapted for use in the ASD classrooms, he was also very aware that the Deaf community especially hungered for access to entertainment films. Rarely did deaf adults patronize theaters since the talkies replaced silent films. Foreign films imported to the United States often did not cater to their interests, and they often did not have plots suitable for children. Boatner studied carefully the student reactions to uncaptioned films, noting their preference for war films, adventure yarns, animated features, and technicolor films. But none of these compared with the pupil reaction to *Crash of Silence* when it was shown with captions.

Boatner noted, "We always envisioned that captioned educational films would be an important area of service, but that project had to wait until later. Our immediate goal was to demonstrate that the production of suitably captioned entertainment films was feasible."[20] This, of course, was a standard first step in any start-up enterprise.

A second important step in starting up a business enterprise was evaluating the market. One important consumer of captioned films would be classroom teachers. In June 1947, Boatner presented a forceful and convincing talk on the need for support to the Conference of Executives of American Schools for the Deaf (CEASD). In expounding on the merits of captioning films, he moved the conference members to establish a committee of three members to develop a plan of operation to provide captioned films for use in school programs. On April 14, 1948, Boatner, O'Connor, and Fanwood (New York School for the Deaf), and Superintendent Charles Bradford met at the Lexington School to outline practical steps. Emerson Romero joined the meeting to help develop guidelines and to share the benefits of his experience as director of the National Film Library for the Deaf.

# 3
## Captioned Films for the Deaf

*The experience led one woman, Mrs. Elsie Durian, to weep with joy—it was the first film she understood in decades.*

—Edmund B. Boatner, "Captioned Films for the Deaf"

With the encouragement of CAID, Edmund Burke Boatner and Clarence D. O'Connor organized Captioned Films for the Deaf in 1949 as a private business financially supported by donations and gifts. The nonprofit organization was housed in an office at ASD. Boatner was president and O'Connor served as vice president.

In October 1950, Boatner presented to CEASD at a meeting held at the Colorado School for the Deaf. He spoke again to the school leaders on the feasibility of establishing a captioned film library for use in schools around the country. He saw great potential benefits for both the deaf child and the deaf adult, not only in furnishing information but also by providing reading practice in a most pleasant form. Boater also described to this group of educational leaders the problems of finding a source of suitable films, arranging captions for the films, discovering a mechanical process to superimpose the captions upon the films, securing funds to provide for a trial program, and working out the handling and distribution of the films, both educational and recreational.[1] The financial and technical challenges seemed formidable, but the project was viewed by everyone as one of great potential and as an educational breakthrough.

J. Pierre Rakow, an inventor who spent years attempting to invent new ways to make the talkies more understandable to deaf people, contributed enthusiastically to the growth of the nascent organization. Rakow and his wife Lillian came to ASD in 1937, partly because the life of an inventor did not satisfy him for long and partly because he became intrigued by his wife's accounts of her teaching experiences at the South Dakota School for the Deaf. Deafened at the age of fourteen, he then attended PS 47 (a day school for the deaf in New York City) for a year and a half before graduating to a public high school, and later New York University. He then became a partner in a typewriter agency in the New York City area, where he gained wide business experience. With Rakow's assistance at ASD, a class in typewriter mechanics was successfully established and three major typewriter manufacturing plants in the area made it possible to place many of the course's graduates in jobs—a great accomplishment in this era.[2] Now, while teaching full-time, the indispensable Rakow oversaw the program's day-to-day operations.

Rakow, who enjoyed the silent films when he was younger, realized that the old method of using subtitles was certainly better than having no access at all, but the subtitles did produce frequent interruption of the action in the film. This caused audiences to tend to grow restive. Now, he began to seek assistance from facilities in Egypt and in Europe. Several of the earliest films were captioned abroad.[3]

Fortunately, within a short time, a technique that involved etching captions directly onto a finished print of the film was devised in Belgium. A negative of the etched print could be made, and this allowed for additional prints to be produced. Titra Film Laboratories in New York was successful in securing a franchise for the new captioning process for the United States, and Rakow was responsible for persuading Titra Film Laboratories to caption the films. Although a major technical turning point in the history of captioned films, financial and marketing barriers remained overwhelming. Rakow also coordinated the efforts of volunteer captioners, aiding them with advice and suggestions. He agreed to do all of this captioning work without remuneration after his regular duties at the school.

CFD, Inc.'s first board of directors comprised business people, community leaders, educators, humanitarians, and even well-known motion picture personalities. Katharine Hepburn was a member as well was Mrs. Spencer (Louise) Tracy, whose son John was deaf. Among the corporators were Spencer Tracy and Helen Keller. Most of the members were not very active, but their names lent prestige to the project. One member, however, Graham H. Anthony, a leading industrialist, would later play a major role in enabling CFD, Inc. to secure federal funding.[4]

The need for funds became an especially pressing problem during the early years of CFD, Inc. Those working on the project were paid for their time or expenses, but the cost of securing films and circulating them was overbearing, and this led O'Connor and Boatner to continuously seek contributions. O'Connor, through his connections, managed to raise a considerable amount of money, and the Junior League of Hartford provided the first substantial financial assistance. Among them was a check of $1,295.29 from the estate of Lester N. Hofheimer of New York, a nice initial deposit when the corporation opened up its bank account. During the year CFD, Inc. was established, the instrumental Junior League members raised a sum for the Connecticut Society for Crippled Children and was looking for their next project. Eighteen organizations were being considered. Boatner later wondered if Katharine Hepburn's sister, Marion Hepburn Grant, a cofounder of the league, may have influenced the decision to select CFD, Inc. With the league's support, the small captioning enterprise was on its way.

The financial commitment made by the Junior League was welcomed, but the funds were not immediately available. To expedite such support, the league first produced a variety show, "The Junior League Follies," to raise support for the captioning efforts. The first gift of $5,000 was followed by one of $2,500. Individual league members also began assisting by actually writing captions for the films. Two members, Frances Beekley and Nancy Callahan, helped to develop captions for the technicolor film *America the Beautiful*, a twenty-five-minute narrated movie created in 1945 by Warner Bros. for $100,000 to sell US savings bonds. In 1951, this film was the first to be open captioned in the United States using the Belgian technique. It was donated by the Department of Treasury and was placed in circulation two years later. It painted a picture of the United States as an industrious society and was directed by six-time Academy Award–winner Gordon Hollingshead. The film began at "your front door, on the street where you live" and spanned the American continent through ranches and farms, national parks, and the bustling cities of the East Coast.

Nothing was more dramatic at this time than the invitational viewing of this film by a group of deaf adults in Hartford. The unique beauty in this film for the deaf viewers was in the *access* to the soundtrack through captions! The experience led one woman, Mrs. Elsie Durian, to weep with joy—it was the first film she understood in decades.

CFD, Inc. then purchased two other films. *The Noose Hangs High*, a comedy by Abbott and Costello, was a gift from ASD. This film, and *Scrooge*, were obtained on August 16, 1951, and copies were circulated in December that year to the New York School for the Deaf and to the Lexington School. Responses were extraordinary among the deaf children, and other titles were soon acquired. Eventually CFD, Inc. obtained, captioned, and rented out a small library of films. Rakow, a shrewd bargainer, deserves the credit for seeking the films.

Handling and distributing the films was also complicated. Boatner knew well that when the film library was developed to where it served many schools, a systematic arrangement would have to be made, but the time for this had not yet arrived.

Feasibility of the enterprise was dependent on too many factors. Due to delays by the mechanical captioning company in New York, Boatner did not get the three new films into circulation until after Christmas in 1951. One of these films was not captioned until late the next spring. It appeared that the captioning laboratory was overwhelmed with other requests not related to the needs of deaf people. Nevertheless, the films were kept in full circulation, completing a total of sixty-seven showings in thirty-two schools.

Boatner was naturally interested in the reception of the early batch of films. In most cases there was much enthusiasm, undoubtedly demonstrating the great need for such a program. The comments on a showing of *Spanish Main*, a feature color film, were summarized in a letter from one principal who wrote to him. This administrator felt that the accessibility of the films could be summed up in one word: "Magnificent."

When sending out the films, CFD, Inc. frequently asked for criticism and feedback. Respondents reported a practically 100 percent agreement that captioned films were better for deaf people and that there was a great need for them. One criticism of the films was that the captions didn't follow the dialogue word for word. This was not possible to fix at the time. There wasn't enough room on the film to write such captions, especially since many deaf viewers could not read them at a rapid rate.

As to the type of film, in the beginning the organization took such films as were available, and these were not always the best films. Of course, they could never expect to please all varieties of tastes as to film selection. Nor did they have experience in the intricacies of captioning films, and allowances had to be made. It was a slow process to produce films of better quality and use the captions to explain the stories more satisfactorily.

CFD, Inc. reported figures from one major film producer, which indicated they would be willing to furnish a 120-film program, complete with shorts, all captioned, at an average cost of $600 for a four year rental per film. Boatner estimated that this would be sufficient to provide sixty schools with a full year's program of forty showings, and if this arrangement could be effected, the rental cost to the schools could be reduced to $10.00 to $12.00 per showing. If this program could be put into effect, he felt they would have accomplished its objective as far as recreational films were concerned, and the library could then devote its attention to educational films and the means of making them available to schools.

The primary need of the new captioning enterprise continued to be financial support to maintain its operations. The cost of the 120 films would be $72,000. Although appearing to be a lot, Boatner reasoned that it would be reasonable if it was divided among sixty schools. This was just one approach. Another approach was to attempt to secure backing from a foundation, and Boatner believed that they had a good chance to secure some favorable consideration. This was the direction they chose to pursue.

## MOTION PICTURE PRODUCERS' CONCERNS ABOUT DISTRIBUTING THE CAPTIONED FILMS

With limited funds, it was necessary to charge modest fees for the distribution of the films—$15 for a black-and-white feature and $20 for the rarer color films. These movies were circulated almost entirely to schools for deaf children. Letters were sent out to inform the schools that a film would be delivered on a particular date, and instructions for passing it on to the next school were also provided. The only complaint the program received was that new films were not sent out often enough.

Film piracy was a concern the motion picture industry faced, and producers did everything possible to guard against it. Consequently, the concern over selling or leasing prints of the better films made it difficult for CFD, Inc. to procure films. This was true despite formal agreements that the films would be shown only in schools for deaf children. In an effort to establish film sources, Boatner seemingly was at odds with the entire film industry. Films were protected by copyright laws, and authors, producers, actors, and local movie exhibitors all had a stake in them.[5] Producers also worried about the films being used on television or in other ways that would cut into box office profits. The companies looked at 16 mm showings in churches, schools, and other organizations as hurting their business. Theaters were their primary source of income. For these reasons, the film companies were especially resistant to the establishment of a program like this group was proposing. One exception was RKO Pictures, which remained a very cooperative studio, and most of the films were obtained from this company.

Third, the technical process of captioning remained a difficult challenge to overcome. Despite the efficiency of the etching technique used by Titra Film Laboratories, translations of the spoken dialogue and other sounds were often unavailable, and were costly to produce. Consequently, the program was forced to accept only films for which the scripts were available. The scripts often arrived in loose-leaf binders, several inches thick, for full-length features, and had to be greatly condensed to be converted into captions. In particular, the length of each caption was calculated and inserted in order to extend only through the scene described and not overlap into the next scene. The problem of timing was so complex that the cooperative RKO had no solutions to offer the program.

Boatner, not one to be frustrated easily, addressed this issue directly with producers. Just prior to leaving one meeting with RKO in December 1951, he asked whether the company exported films to foreign countries. The answer was affirmative and he asked, "Are the captions written in English before being translated into the foreign language?" RKO explained that they did work out captions in English prior to translating them. This was another technical breakthrough, one that simplified the caption timing problem for CFD, Inc. For the time being, captions would be taken directly from the scripts written for foreign films. The captioner would know the exact length of the insertion and had only to maintain the same syllabic count. In terms of the timing issue, a skilled captioner could now arrange suitable captions without even reviewing the film.[6]

A fourth challenge the program faced during these early years of operation was that the language level of the captions needed to be suitable for deaf viewers, whose reading skills often were poorer than those of hearing viewers of the same age. This required rewording of the captions by someone experienced in the language limitations of deaf people. This issue would remain a major concern for years to come.

Regardless of these difficulties, the establishment of this small private organization sent excitement throughout the American Deaf community, and it wasn't long before Titra Film Laboratories became even more overwhelmed with orders. The company nevertheless continued to faithfully caption the program's films, despite some delays. Later, it would farm out some of the captioning assignments to companies in Belgium and other countries that had the proper equipment. The work was comparable to that produced stateside, but an average of three months was required for the preparation and shipment of each film.

## FORMALIZING THE ENTERPRISE

Entrepreneurs often face a slow process of putting everything on a legal footing. Many businesses start up their operations while lacking

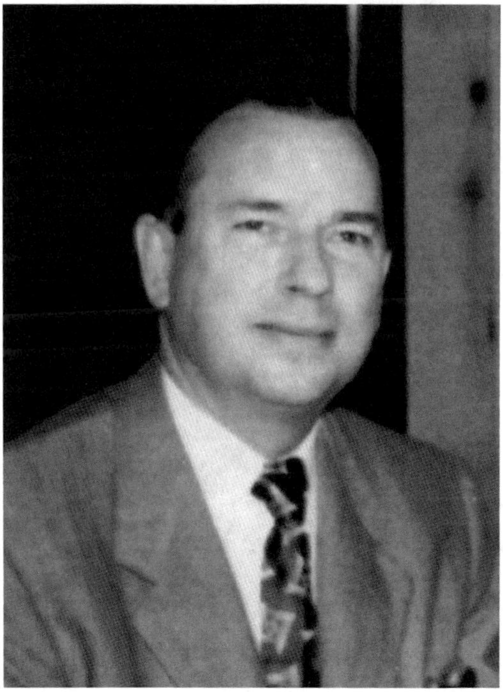

Edmund B. Boatner, superintendent of ASD in Hartford, Connecticut, was one of the principal organizers of CFD in 1949. Boatner served as president. Image courtesy of the Described and Captioned Media Program.

registration with the required authorities. It is the process of formalization that is one of the most important problems faced by start-ups.[7] On March 21, 1955, the organization was incorporated as CFD, Inc. under the laws of the State of Connecticut with Boatner as president, O'Connor as vice president, and Rakow as staff coordinator. Among the directors were Leonard M. Elstad, president of Gallaudet College; Graham H. Anthony, industrialist; Alfred C. Fuller, chairman of the board, Fuller Brush Company; and James H. Galloway, CAID president. Among the corporators were Byron B. Burnes, president of NAD; Katharine Hepburn; Spencer Tracy; and Mrs. Louise Tracy.

By this time, the program had acquired seventeen prints. As the films began to show wear, Boatner realized that wear and tear were fast outrunning the capacity of the corporation to finance new purchases. Expansion of the program was becoming all but impossible as a private venture.

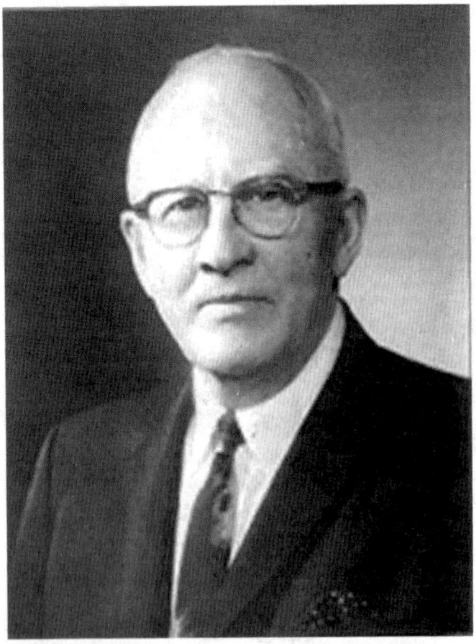

Clarence D. O'Connor, director of the Lexington School for the Deaf in New York City, served as vice president of CFD in 1949. Image courtesy of Described and Captioned Media Program.

# 4

# Obtaining Federal Support: Public Law 85-905

*I am sure these films will enrich the lives of our fellow citizens afflicted with deafness—and through them, will benefit the national community.*

—President Dwight Eisenhower, letter to Edmund B. Boatner

As time passed, it became apparent that CFD, Inc.'s goal of making captioned films available to deaf people throughout the country was too vast an undertaking to be achieved by a single small private organization. Edmund Burke Boatner and Clarence D. O'Connor made the decision to seek an act of Congress that would provide for the production and circulation of captioned films for America's deaf population. At first, they ignored warnings that it was unlikely that a bill applying to a single disability group could get through Congress. They knew that the US government was already funding a Talking Books for the Blind program and they reasoned that the Library of Congress would be a logical place for the film program.

## A NATIONAL LIBRARY OF CAPTIONED FILMS ENVISIONED

By this time in history, the "library for the blind" usually referred to regional libraries that were the primary delivery source of books and magazines for blind Americans. In addition, the National Library Service for the Blind and Physically Handicapped of the Library of Congress in Washington, DC, was a federal program that provided materials and services to the regional libraries. This included books (discs, cassettes, and braille), magazines, equipment (talking book machines, cassette machines, and their accessories), catalogs, and support and informational services. The National Library Service did not provide any direct funding to the regional libraries. Any person who was legally blind or who had visual disabilities that prevented the reading of standard print material was eligible for service from a regional library.[1]

In 1897, John Russell Young, the Librarian of Congress, created a reading room specifically for blind people with about 500 books and music items in raised characters.[2] The passage in 1904 of legislation that allowed embossed reading material for the blind to be sent through the US mail postage free was the first time that the federal government acknowledged that blind people did read.

In 1913, Congress mandated that one copy of each book in raised characters made for educational purposes under government subsidy by the American Printing House for the Blind in Louisville, Kentucky, was to be deposited in the Library of Congress. Other materials were acquired by gifts and purchases. Expanded services included instruction for those desiring to learn to read by touch. The library also developed plans for exhibiting products made by blind persons.

By 1930, there were at least eighteen libraries serving blind people without federal aid. However, federal funds were needed to alleviate the financial burden, and the American Library Association asked the American Foundation for the Blind to help address the problem. A survey led to the request for federal legislation and, in 1931, the Pratt-Smoot Bill (PL 71-787) was passed. This bill called for the creation of a program within the Library of Congress called the Division for the Blind, which had an annual appropriation of $100,000 (the ceiling on appropriation at that time) to produce and circulate embossed reading material to blind adults.[3]

Funding milestones included President Roosevelt allotting $211,500 in 1935 to the Works Progress Administration for the construction of 5,000 talking book machines "for the purpose of enabling the blind to use the books now provided by the Library of Congress."[4] In 1947, the Library of Congress made the first purchase of talking book machines. The following year the National Bureau of Standards developed specifications for a low-cost, reliable talking book machine. In 1952, the word "adult" was deleted from enabling legislation, making children eligible for services. In addition, consideration was given to braille printing houses adopting *English Braille, American Edition,* as the standard for braille produced in the United States. Books on open-reel tape were circulated to readers as a supplement to the talking book program.

Boatner and O'Connor saw a major difference between the programs that might be an obstacle for them. The service for blind people was based on an already-existing distribution system principally through public libraries. The law did not call for the creation of a distribution system, but only for materials to be distributed. However, the federal grant did include money for research, which resulted in the talking book program. Such an ambitious federally funded program seemed to be a good prototype for the efforts to caption films. In addition, there were obvious similarities in the program for blind people and the one being envisioned for deaf people. They were both designed to assist a sharply defined audience controlled in size, and they were both designed to develop materials that stress one sense over the other. They both translated one type of record into another type of record. Both had similar opportunities and problems; both were subject to federal regulation. Both were required to pay for the materials. Neither was a commercially profitable venture, which meant both needed federal subsidy.[5]

Boatner and O'Connor knew that for blind people, libraries were distributing the materials because the materials were books in a different form. For deaf persons, however, film distribution offered different problems. And there was a difference in numbers of duplicate copies needed, because groups would use the captioned films whereas individuals used the books. Circulating the films to schools required more voluntary work. Boatner was the administrator and personally wrote letters to school officials, many whom he knew in CAID. His secretary, Dorothy Spencer, handled the details of shipping and billing. Still dependent on gifts, the corporation considered it a "windfall" to receive a $650 check through the generosity of a local beneficiary, William A. Mauser.

Boatner and O'Connor discussed the issues with Verner W. Clapp at the Library of Congress, and he gave them warm encouragement.[6] Clapp had a long history of supporting blind people. As early as 1941 he gave a speech titled "Some Problems in Library Service for the Blind."[7] The idea of having the Library of Congress as CFD, Inc.'s home was worth exploring.

Graham H. Anthony, the industrialist who was a member of the CFD, Inc. board as well as a director of ASD, was serving on the advisory committee for the Vocational Rehabilitation Division of the Department of Health, Education, and Welfare (HEW). Mary Elizabeth Switzer, the director of the division since 1950, was one of the most energetic and inspirational people in the field. She became best known for her work on the 1954 Vocational Rehabilitation Act, which had a tremendous impact on rehabilitation services for people with disabilities.

When Anthony first visited ASD, he delighted the printing instructor by fingerspelling to him. As a young boy, Anthony roamed the hills of North Carolina, hunting and fishing with two deaf friends who attended the North Carolina School for the Deaf. Through those friendships, he became interested in the welfare of deaf people throughout his lifetime. As Boatner and O'Connor were struggling to get the captioning project to move forward, it was Anthony who suggested an attempt be made to get an act through Congress to make CFD, Inc. a federal program.

In 1956, with great excitement generated over the films then in circulation, Anthony persuaded Senator William Purtell (R-CT) to introduce a bill in the US Senate to establish an agency to procure, caption, and

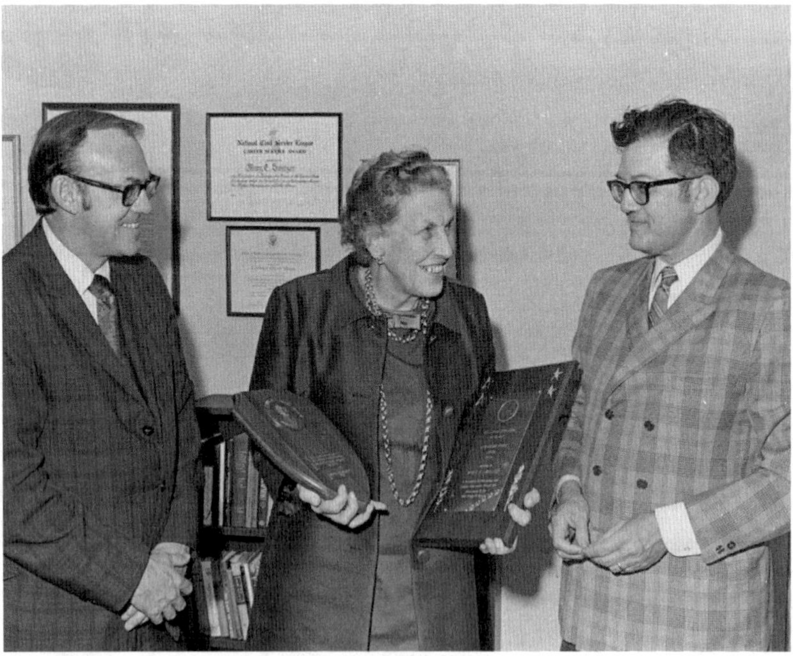

Mac Norwood (left), with Mary E. Switzer and Frederick C. Schreiber, executive director of NAD. Image courtesy of Ernest E. Hairston.

distribute suitable films. When Purtell, who was quick to see the humanitarian service he could render, enthusiastically introduced Senate Bill 3558 on March 26, 1956, he called for an authorization of $250,000 to establish an agency to carry out these responsibilities. The agency was to be operated by the Library of Congress, as Boatner and O'Connor hoped. Purtell was able to secure the cosponsorship of forty of his senatorial colleagues, and there was great confidence for the bill's passage.

Senator Purtell was moved by a visit to ASD. The visit was a constant inspiration, and in one of his pleas in the Senate he told fellow legislators of an experiment he conducted in his home after this visit. He turned off the audio on his television set and attempted to follow the story. His failure to understand the show increased his determination to push the measure in the Senate.[8] Purtell's bill, if passed, would enrich the entertainment, cultural, and educational experiences for the deaf population, and end once and for all CFD, Inc.'s constant financial woes. A similar bill was introduced in the House by Representative Albert P. Morano (R-CT).

In the meantime, the demand for film services by deaf people was becoming insatiable, and CFD, Inc. was placing newer and better films in circulation as quickly as the laboratories could process them and finances would permit. RKO was no longer the sole source of major films when Columbia Pictures and United Artists joined the fold. Columbia released captioned versions of *Military Academy*, *The Pathfinder*, and *Rogues of Sherwood Forest*. Walt Disney's *Treasure Island* was the fourth film. The program felt it was a coup to establish Disney as a major source of films. Before long came *Snow White and the Seven Dwarfs* and *Pinocchio*. It wasn't clear how the film program would be able to tolerate Congress's notoriously slow pace.

## LIBRARY OF CONGRESS ARGUES AGAINST TAKING RESPONSIBILITY FOR CAPTIONED FILMS

After the Christmas recess, word came from Washington that things were not going well. The Library of Congress wanted nothing to do with the new Captioned Films for the Deaf program being proposed. Another visit to Washington was in order, and O'Connor and Boatner soon learned that the rumors were all too true. Verner W. Clapp was no longer with the Library of Congress, having left to take a position with the Ford Foundation.

On January 4, 1957, Boatner, O'Connor, and Patricia Cory, a librarian from the Lexington School, met with the new Librarian of Congress, Lawrence Quincy Mumford, and members of his staff. According to Boatner, Mumford's attitude was at a complete variance with the attitude that Clapp had exhibited earlier. Where Clapp showed keen interest in the project and gave them warm encouragement, "Mumford stated bluntly that he did not feel that Captioned Films for the Deaf came within the province of the Library."[9] When it was pointed out to Mumford that the library was serving blind people, he replied that if it were up to him, he would dissociate the library from that activity as well. Mumford did add, "However, we are the servant of the Congress, and will do whatever it directs us to do."[10]

O'Connor and Boatner were angry. By law, the Library of Congress was receiving two prints of every film copyrighted in the United States and had what was the largest collection of films in the world. Boatner and O'Connor then met with Boyce Williams, the deaf consultant from the Office of Vocational Rehabilitation in HEW. Further discussion with Mary E. Switzer brought agreement that "it was not a good idea to have the film program managed by an agency that did not want it."[11] The bill was reworked to place it under the direction of Williams's office, and he agreed to do the work necessary to build up a case for federal legislation.

On April 16, 1957, Senator Purtell introduced Senate Bill 1889. Unlike the previous bill, the new bill stipulated that the program would be administered under the direction of HEW. The bill was referred to the Senate Committee on Labor and Public Works chaired by Senator Joseph Lister Hill (D-AL). Although he was not particularly aware of the importance of this bill, his attitude most definitely changed when he received a large batch of personal letters from nearly every deaf child at the Alabama School for the Deaf. The letters moved him considerably. Soon thereafter, Representative Morano reintroduced the bill in the House, where it was referred to the House Committee on Education and Labor.

## PUBLIC LAW 85-905: A WATERSHED

The new version of the bill passed the Senate on August 23, 1957, but when it reached the House, it became buried under a logjam in the Committee on Health, Education, Labor, and Pensions. The session of Congress was reaching its end, and although no one was against the bill, its chances of clearing that committee seemed impossible. The committee's chairman decided not to convene the committee before the end of the session because a major bill, to which he was opposed and which he was determined would not be voted on, was coming up. The only way he could be sure of the outcome was not to convene the committee at all.[12]

After this delay, the bill was returned to the House Committee on Education and Labor several times. But after Representative George S. McGovern (D-SD) provided supportive remarks on August 7, 1958, the bill was passed by the House on August 15.[13] Then, on August 21, 1958, Captioned Films for the Deaf bill became PL 85-905. It was signed into law by President Dwight D. Eisenhower on September 2, 1958.

The purposes, as set forth in the law, were threefold:

> (1) to bring to deaf persons understanding and appreciation of those films which play an important part in the general and cultural advancement of hearing persons; (2) to provide, through these films, enriched educational and cultural experiences through which deaf persons can be brought into better touch with realities of their environment; and (3) to provide a wholesome and rewarding experience which deaf persons may share together.[14]

Under the provisions of PL 85-905, Captioned Films for the Deaf was authorized to:

> (1) Acquire films (or rights thereto) by purchase, lease, or gift; (2) Provide for the captioning of films; (3) Provide for distribution of captioned films through State schools for the deaf, Deaf clubs, and such other agencies as the Secretary may deem appropriate to serve as local or regional centers for such distribution; (4) Make use, consistent with the purposes of this Act, of films made available to the Library of Congress under copyright laws; (5) Utilize the facilities and services of other governmental agencies; and (6) Accept gifts, contributions and voluntary and uncompensated services of individuals and organizations.[15]

The dissolution of Captioned Films for the Deaf, Inc. in Hartford on February 29, 1960, included the transfer of its motion picture library to the Office of Education, the agency designated by Secretary Arthur S. Flemming of HEW to administer the service. After this law was passed, CFD, Inc. became simply CFD.

Since its inception, this federal program acquired, captioned, and distributed hundreds of film treasures that deaf Americans would have otherwise been unable to enjoy either in schools or theaters.

In a fitting tribute, President Dwight Eisenhower wrote to Edmund B. Boatner. Impressed with the pioneering work of Captioned Films for the Deaf, Inc.,[16] Eisenhower expressed pleasure in taking part in the enactment of the legislation that would allow this work to continue. "The collection of films which your group gave the Government provides a substantial beginning for the Federal program in this field. This gift, together with the enthusiasm and experience which accompanied it, are most appreciated. I am sure these films will enrich the lives of our fellow citizens afflicted with deafness—and through them, will benefit the national community."[17]

# 5

## The Captioning Program Begins

*I fully believe that the legacy of captions is the most positive thing that has happened in education of the deaf in this century.*

—Frank B. Withrow, letter to John A. Gough, June 11, 1992

After the passage of the Captioned Films Act of 1958 (PL 85-905), the Commissioner of the US Department of Education Lawrence G. Derthick quickly realized a dynamic leader was needed to spearhead the new CFD. He called a preliminary conference on November 21, 1958, in Washington, DC, to consider and discuss questions and operations with administrators of school programs serving deaf children, representatives of organizations serving deaf people, and several people from the motion picture industry and the Offices of Education and Vocational Rehabilitation. Derthick conducted a thorough search for a capable administrator based on recommendations from numerous organizations. When the search turned to Gallaudet College, he found John A. Gough. On November 13, 1959, the Department of Health, Education, and Welfare (HEW) announced the appointment of Gough to direct CFD.

Gough was born on a farm near Muncie, Indiana, on January 18, 1907. He pursued undergraduate studies at Ball State. His first professional contact with deaf people occurred upon his graduation from Ball State Teachers College in 1927, when his uncle, Oscar M. Pittenger, the superintendent of the Indiana School for the Deaf (1919–1935), asked him to serve as part-time teacher of language and social studies and as basketball coach. Coincidentally, it was in 1927 that Pittenger wrote in the school's *Eighty-Third Annual Report* about the effective use of "Motion Pictures."[1]

Realizing the need for more formal training for themselves, Gough and his recent bride, Harriet Sterling Nay of Indianapolis, applied for fellowships in the teacher preparation program (then called the Normal Department) at Gallaudet College, where both completed their studies in 1932. This was the beginning of a professional partnership that immeasurably benefited at least three generations of deaf children.

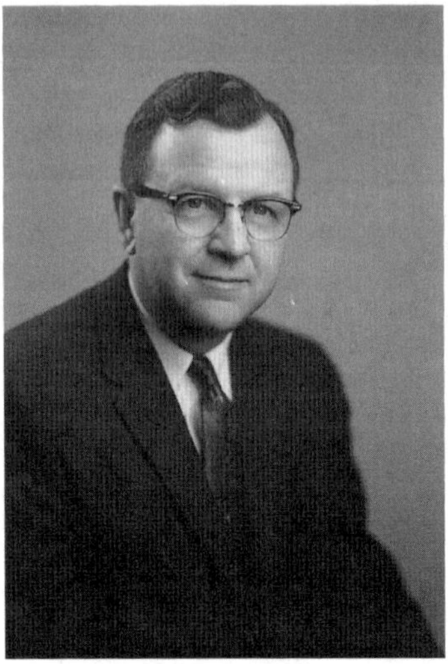

John A. Gough, first chief of Captioned Films for the Deaf. Image courtesy of Susan Gough Updegraff.

He returned to Indiana as a full-time teacher and found time to earn his master's degree in school administration at Indiana University. In 1934, he joined the faculty of the Kendall School for the Deaf on the campus of Gallaudet and remained as an English and mathematics teacher until 1937, when he became principal of the Georgia School for the Deaf in Cave Spring. Subsequently, he served as supervising teacher at the Missouri School for the Deaf in Fulton, and he was appointed superintendent of the Oklahoma School for the Deaf in 1940.

Gough also attended Johns Hopkins University and the University of Oklahoma. Although he completed all the course requirements for a doctoral degree, he did not finish his dissertation because he received a lucrative offer to enter private business in Washington State in 1946. Although no longer directly connected with the education of deaf children, he served as a consultant to the State Department of Vocational Rehabilitation as an interpreter, a teacher of American Sign Language in adult education programs, and an employer of deaf persons. In 1957, he returned to the field of education as chair of the Department of Education at Gallaudet and principal of the Kendall School. His foresightedness and leadership in upgrading and broadening the teacher training program at Gallaudet made him the logical choice to head the newly formed CFD program.

Gough knew from the start that making motion pictures accessible to deaf and hard of hearing people portended a breakthrough of epic proportions. Indeed, he was honored to be asked to pioneer the program development on the federal level.

The Deaf community was both grateful and excited about Congress awarding the funding to CFD to begin an organized approach to provide captioned films for their enjoyment. Gough, however, faced an immediate challenge when Congress adjourned before funds could be appropriated to CFD. During this period, both Congress and the Office of Education were giving high priority to the implementation of the National Defense Education Act (NDEA). In response to the Soviet launch of the satellite *Sputnik* and the subsequent acceleration of the space race, the NDEA provided federal funding for trained manpower through fellowships and loans to students. The legislation bolstered education in the areas of science, mathematics, and modern foreign languages. Making sure that this legislation was carried out effectively necessarily occupied a great deal of time and attention in the Office of Education.

In addition, during the early stages of federal oversight of the CFD, copyright concerns surfaced regarding the application of a provision contained in PL 85-905 that allowed the use of films made available to the Library of Congress. To allay these fears, the Office of Education took action to ensure that no infringements would be made on the rights of copyright owners. The Library of Congress' copyrighted collection of works would not be used, and rights to use films for captioning would have to be negotiated with the producers (i.e., the copyright owners).[2]

After these issues were clarified and the NDEA launched, the Office of Education could turn its attention to implementing the CFD program. The $78,855 secured to begin CFD included $50,000 earmarked for acquisition of titles. The rate of future expansion was dependent to a great extent on the demand for the films. In addition, there were many other projects sponsored by the Office of Education, and the competition for funds and services was great.[3] This groundbreaking law was a watershed mark in the history of captioning but was just a beginning in the long venture for federally funded access to the spoken word.

## THE FIRST EFFORT TO TRAIN CAPTIONERS

CFD was the only player in the captioning movement at this time. It continued to struggle to circulate a few entertainment movies it had acquired and captioned. Deaf people in social clubs and state schools were charged a very small fee to cover postage and other expenses. From the beginning the insatiable demand far exceeded the supply, and it became apparent that more money was needed for CFD.

Importantly, Gough realized that PL 85-905 only allowed for captioning of feature films, and he saw that it was necessary to amend the newly passed law to include educational films. Although public schools had been using films for decades, students who were deaf still had no access to this valuable teaching tool. Moreover, prior to the enactment of PL 85-905 on August 20, 1969, there was very little research done in the use of film and television for deaf learners.

Training in captioning the films began in December 1959. Although CFD's technical consultant, J. Pierre Rakow, had some experience in

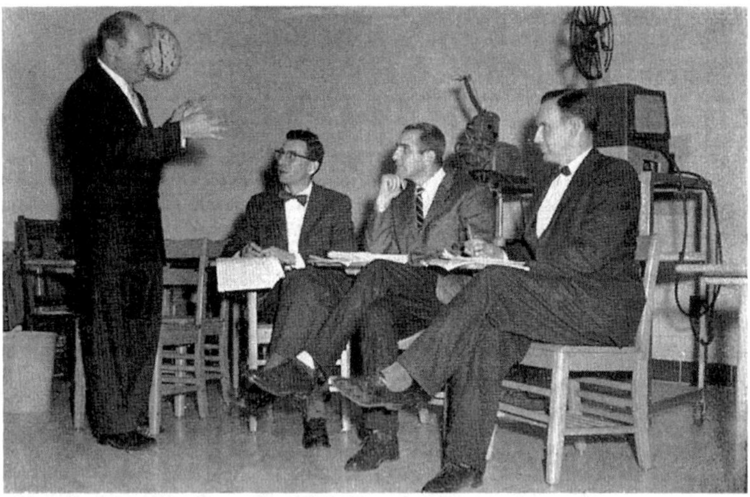

J. Pierre Rakow of Hartford, Connecticut, a consultant for the US Office of Education's CFD, describes some of the techniques of film captioning to Stanley Benowitz, left, Robert F. Panara, center, and Leon Auerbach, right, all of Gallaudet College. Image courtesy of Wallace Memorial Library at Rochester Institute of Technology.

captioning, for the most part such training largely had gone unexplored to date, and now it presented many complex issues. As a first step, Rakow traveled from Hartford to Gallaudet College to spend three days sharing his knowledge and skills with deaf Gallaudet instructors Stanley Benowitz, Robert F. Panara, and Leon Auerbach. They were the first three captioners in CFD. Later, a fourth teacher, Terrance J. O'Rourke, was also trained to be a captioner. Rakow was also captioning films through a private company he had established, Superior Films, from which he also consulted for the CFD program for several years.[4] These efforts led *The Silent Worker* to claim that the CFD program was "not only for, but also by the deaf."[5]

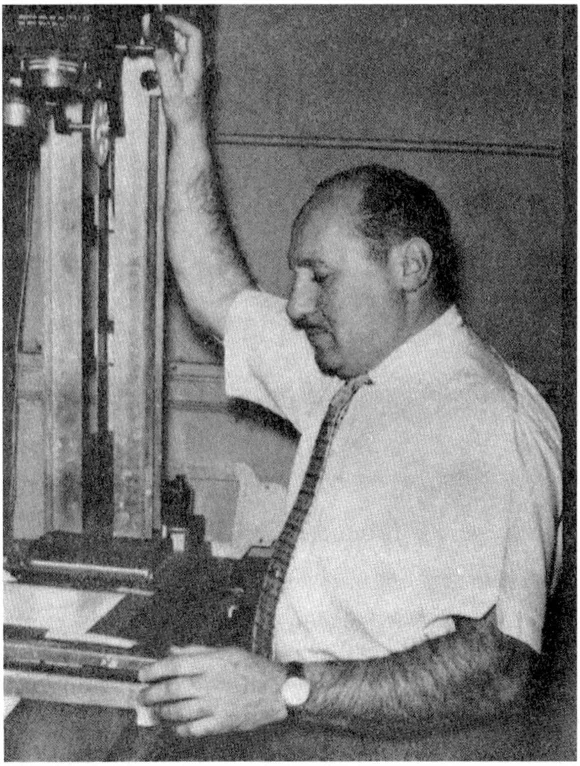

J. Pierre Rakow, a deaf pioneer in captioning films, prepared more than thirty films before 1960. Image courtesy of Wallace Memorial Library at Rochester Institute of Technology.

In particular, learning to caption at this time in history required a measure of technical knowledge about the mechanics of a film projector. It was not a simple process of just adding words to certain frames of film. Rakow taught the Gallaudet instructors how film passed through a projector at the rate of twenty-four frames (or single pictures) per second. Thus, the film moved, and the pictures seemed to move, but the captions must stand still in order to be read. To accomplish this, the same words must appear in exactly the same position on a series of pictures that make up one scene.[6]

In addition to making captions stand still, there was the problem of keeping them out of the way so that they did not overlap with the picture in the frame. The captions also had to be inserted so that they appeared and disappeared at the right time. Most scenes in a movie last only a few seconds and the captions had to be quite short. This was done by condensing the words of the speakers, giving only enough of what they said so that the deaf viewer could get the sense of what was happening. Benowitz, Panara, and Auerbach were experienced educators familiar with the language development challenges of deaf children and adults.[7] They were fortunate to have Rakow's guidance. At the time, he was the primary expert in the United States interested in captioning films for deaf people.

After a few years the enthusiasm over the CFD's captioning efforts grew quickly, and there remained much to be done to expand the national library of entertainment films. The films were still being captioned by the instructors at Gallaudet College. These deaf captioners were unable to hear the script and were guided by a special "spotting script" on which dialogue was measured according to the number of frames. Although the situation was far from ideal, these pioneers were motivated by the belief that deaf children had a better chance to learn through such access.

## DEPARTMENT OF AGRICULTURE ASSISTS IN DISTRIBUTING FILMS

By December 1959, arrangements were made with the Motion Picture Division of the Department of Agriculture to handle the booking and shipping of films to be distributed by CFD. Gough's office was paying for this service. The first actual booking under the auspices of the US government was received from a New York parochial school where the children enjoyed a showing of the Disney picture *Dumbo* as a pre-Christmas treat. In regard to this first CFD booking, Gough wrote to Boatner, noting the historic importance of a letter received from Mother Mary of St. Mildred SMR Convent of Mary Reparatrix in New York. He thanked Boatner for setting up the first film booking under CFD, stating that "we are about to get under way. I hope that the program may grow to be an ever-increasing help to the deaf."[8] This letter represented the culmination of much work by the dedicated educators Boatner and O'Connor, the end of a major phase in the history of captioned films for deaf people, and the beginning of a new frontier where dreams might be played out with federal support.

## EXPANDING THE FILM LIBRARY

By the time of Gough's appointment, CFD had received the twenty-nine feature-length films from Boatner, five of which were Walt Disney films. CFD also had several short subjects and was already circulating all of these films among schools serving deaf children. These schools had a long wait time to gain access to the few films that were made available.

Consequently, Gough immediately began negotiating with film producers to develop criteria for acquisition of more films. He also sought to determine the most favorable captioning techniques, and he personally attended the training workshops led by Rakow at Gallaudet College. Although the films in the library were primarily for entertainment, he quickly realized the great possibilities in captioned films as a means of breaking the language barrier that had been limiting the scholastic achievement of deaf students. He thus launched a campaign to broaden the scope of CFD, an appropriate step considering that funding was coming from the US Office of Education. In particular, Gough began stressing that the learning experiences of deaf students were almost entirely visual and that the development of specialized materials for use by teachers, parents, social workers, rehabilitation counselors, psychologists and others would be a means of providing the all-important language skills so vital to learning. The program Gough inherited from Boatner and O'Connor seemed to him to be a natural starting point for addressing the language development issue with more emphasis on educational films and other visual media.

On October 9, 1959, before he was officially appointed as head of the CFD program, Gough held a conference with five deaf professionals and six who were hearing.[9] Gough led the discussions with his supervisor, Seerley Reid, head of the visual education section of the US Office of Education. It was this group that decided that the films were to be categorized as educational, recreational, or cultural. The group also discussed regulations governing film service, procedures for the care and return of films, and the establishment of an evaluation committee to help select films. The deaf writer Max Friedman summarized this work with an optimistic report on how the forthcoming films would include topics covering adventure, biography, sports, travel, history, and comedy. Friedman concluded, however, with an important cautionary note that as time goes by, the library should include a well-chosen selection of films in every category suitable for every section of the deaf population, both young and old.[10]

Enthusiastically approaching his adventure as head of CFD in 1960, Gough sat down and developed a "dream book." In *A Forward Look*, he summarized his initial perspectives on combining and centralizing resources. From the start, more funding was a priority. He wrote, "If a total job is to be done in supplying educational media for the use of those with hearing impairment, it is apparent that the concept of media at the federal level must be broadened and that federal financial support must be substantially increased. It would seem to be sound procedure to consolidate administration of all these media in a single dissemination service."[11]

Importantly, John Gough was predicting that a much broader media service for deaf people might actually supersede CFD. With this dream book began a revolution in media services in the education of deaf students that had its roots in the initial funding for captioned films.

In addition to the inadequate funding making acquisition of films difficult, film companies were concerned about the copyright issue, that is, the reference in the authorizing legislation to making use of films in the Library of Congress. Although this matter remained a sensitive one, Gough was nevertheless very motivated by the supportive letters coming in from all over the country. In *The California News*, Hugo F. Schuhoff, superintendent of the California School for the Deaf at Berkeley, praised the valuable educational and recreational services now afforded by

Captioned Films for the Deaf. He stressed the tremendous possibilities in educational programs as well as those that met the recreational needs of deaf people.[12]

During the 1960s, the articles, letters, and copies of school newsletters that poured into the CFD office offered many complimentary remarks about the captioned films services being created. Importantly, the deaf children's remarks emphasized both the entertainment side as well as understanding and learning.

At the Indiana School for the Deaf, where boys from the Kentucky School for the Deaf were visiting, Walt Disney's *The Desert* was shown. "Having viewed this film, both without the captions, and with them," the latter school summarized in a newspaper, *The Kentucky Standard*, "we can truthfully say the pleasure was more than doubled by the aid of the captions. Our sincere congratulations go to those who have brought captioning films into being."[13]

In Columbus, the Ohio School for the Deaf students watched *Gentleman Jim*, a film about James J. Corbett, the master boxer of the 1890s. Before the film was shown, the boys and girls had taken their seats in the gymnasium. There was an air of great expectancy, and the students were wonderfully satisfied by what followed on the screen. "Never have pupils appeared more delighted following a film presentation. A movie can truly be enjoyed, now. . . . These films represent a high quality of captions. They showed vitality and carried the story clearly to the deaf in the perfect mood of the particular scenes being shown on the screen."[14]

In Philadelphia, Vinita Seward wrote another motivating note. She was the only hearing person at a recent social, where a movie with captions was shown. "Due to some miscalculation or lack of understanding of how long the film would be run, it was started about midnight. It was fifteen minutes before 3 A.M. when it finished, but thirty-two deaf persons sat breathless to watch this marvelous film. I think that is a marvelous testimonial to the worth of your work."[15]

In the Seattle area, where the interest in captioned films was overwhelming, many deaf children and adults remarked that they had seen some of the films years earlier but did not understand them, how the films are a source of both amusement and instruction, and how the young deaf children sit entranced through a whole showing. One deaf child wrote that she cried while watching *So Dear to My Heart* because the film was so good. It was with this sense of promise that Gough planned a national meeting to begin to translate his dream book into reality.

## THE JUNE 1960 CONFERENCE: ESTABLISHING PRIORITIES FOR CAPTIONED FILMS

There were many other issues associated with captioning films with which Gough wrestled in the short time he had been appointed head of CFD. Which educational films should be acquired? What, if any, responsibility would the office have for achieving good and effective use of the films? What types of groups should be recognized as eligible for services? What responsibility does the US Office of Education have to help groups in following requirements and regulations? Should the service attempt to provide equipment?

In terms of building the collection of films and determining its range, should the film library include a whole range of films, including features, documentaries, short subjects, and teaching films? About Westerns and musicals? Both fiction and nonfiction films? Should filmstrips and transparencies also be collected and provided to schools?

CFD would need to make decisions about the subjects covered, age levels served, policies to be adopted with respect to captioning religious films, and whether foreign films should be acquired.

In addition, regarding the captions, they would need to take into consideration the language level used on feature films, on adult documentary films, and on children's films. A larger cadre of captioners would need to be trained on methods to control vocabulary and sentence structure, standards for evaluating captions, and adherence to the soundtrack. Should there be full coverage (i.e., "verbatim" captioning) of the soundtrack? At what speed should the captions appear on the film? How should the captioner handle the sound effects? This is just a small sample of the questions that challenged Gough. These would be the focus of a national conference held on June 8, 1960. This conference was coordinated by Patricia Cory from the Lexington School for the Deaf, a librarian and visual education director experienced with the issues of teaching language and, especially, reading. Cory was the author of *School Library Services for Deaf Children*, in which she advocated that motion pictures should be sought out in all possible subject fields and for all age levels. She stressed that special consideration be given to issues such as whether the film is well organized, focusing directly on its subject, and is factually accurate.

There were thirty-seven representatives from the Deaf community and deaf educators in attendance, as well as representatives from several different specialties within the field of film communication.[16] The conference was sponsored by the US Office of Education in cooperation with the Lexington School for the Deaf. Clarence D. O'Connor and Seerley Reid began by greeting the participants on behalf of their respective agencies. Edmund B. Boatner was there as well.

Julien Bryan, executive director of the International Film Foundation, spoke about the use and value of films for cultural enrichment and for information. He gave a demonstration showing the film *Japanese Family* to a combined group of deaf children from the Lexington School and hearing children from the Hunter College Elementary School. Margaret Wood, the deaf children's teacher, assisted by Bryan, and Cory sitting in a spotlight next to the screen, provided an interpretation for the motion picture using the oral approach (speechreading without sign language) employed in the Lexington School at the time. The discussion focused on what they had learned about Japan from the film. Cory pointed out that even though neither group of children had previously seen the film, an enormous amount of "preteaching" had to be done to prepare the deaf children to understand what they would see. She noted that she had prepared an outline of the film and a vocabulary list, which she provided to Wood, who had spent two class periods pre-teaching the children. In those classes, Wood explained the visual images on the screen but she made no attempt to follow the language and incidental background information being spoken by the film's narrator. The narrator's language seemed too difficult for deaf children, and they could not assimilate all the background information in such a short time.

At the conference, the potential impact of CFD's captioning efforts on the education of deaf students was becoming increasingly apparent. The role of media in the classroom took on new dimensions during the two days, accompanied by the realization of the need for better training for teachers who may wish to use films, captioned or not captioned.

Thus, in terms of educational films alone, CFD was ushering in a revolutionary movement in the education of deaf students.

## THE NEED FOR A DEAF PERSPECTIVE

At the conference on captioning access, Gough felt it was critical to have the perspectives of deaf people—both consumers of film media and leaders in the Deaf community. He had included deaf professionals in the earlier smaller conferences as well. Now, in June 1960, Byron B. Burnes, president of the NAD, was invited to discuss what deaf adults themselves were thinking about in terms of the scope of CFD. Formal educational emphases momentarily put aside, Burnes described the great joy felt among deaf people when the bill for captioned films became a law in 1958, and he recalled how they had written letters to their congressmen urging them to support the bill. Like the writer Max Friedman had done, Burnes stressed the importance of adult deaf people being consulted on a continual basis regarding developments by CFD. The selection of films should reflect adult interests and desires, not just those of the children. He candidly told the conference participants that deaf consumers initially looked askance at the placement of CFD in the Office of Education, where there was (at that time) no expert on deaf consumers or school learners. He described what the deaf consumers believed to be a broad and comprehensive program, one which would "... supply as many as five hundred clubs and other local groups each month and approximately fifteen to twenty conventions each summer, in addition to its supply of films for school use."[17] Burnes also noted that with the present funds, a program of such magnitude could hardly be expected, and he added that deaf people were appreciative and were eager to assist in any way possible.

## A BROAD AGENDA

Other topics covered at the 1960 conference included availability and "captionability" as important factors in the selection of films. In terms of costs, it was estimated that a feature film (usually nine reels) from a studio cost about $600 per copy. Four prints of each title purchased at that time cost $2,400. Captioning cost $1,300 for a total of $3,700.

Prices varied with the age of the film and there was a great difference in cost between color and black-and-white prints. Gough estimated an average cost for four prints of one film might be between $4,000 and $5,000.

The conference participants delved into more depth with regard to the language and reading issues faced by deaf students. Powrie V. Doctor of Gallaudet College led a panel discussion focusing on the desirable language level and readability of captions, vocabulary, speed of the captions, conceptualization, and use of verbs. Boatner pointed out that many deaf children go through life seeing situations that they do not understand but that they accept without questioning because they do not have clues to help them ask questions. He voiced the hope that films with readable captions might bring about a change and help deaf students to question, and thus to understand, what they see. On battery tests, reading scores always came in very low, revealing that the tool the deaf child needs the most is the weakest. Increasing vocabulary and improving reading should be goals of educators using captioned films.

In the early experience in captioning films at the American School for the Deaf, they found they could use about eight words in a caption, using three lines without obscuring the visual image. On a feature-length picture this totals about 4,800 words or about sixteen double-spaced typewritten pages to be read by the deaf persons seeing the film. Some researchers had questioned the wisdom of introducing more than one new word in thirty-seven, and it was clear to the educators and captioners that more research was needed on the difficulty of deducing meaning from the context. Nevertheless, Boatner believed that captioned films could be a painless way of improving reading skills and that captions could definitely improve the viewer's receptive vocabulary.

The conference participants also discussed the difficulty of establishing grade levels for captioned films. For example, technical films may require technical language, and the grade level of the language should be changed to suit the type of film. Boyce Williams, a deaf participant at the conference, expressed the hope that the language used would challenge deaf students to make inferences from what they read in the captions. Other topics covered in the discussions included motivation as a factor in learning through captioned films, increasing independent reading after school hours, the challenge of multiple meanings of words, abstract words that cannot be explained in terms of concrete objects, and the use of idioms.

One thing clearly stood out during the conference—captioning should be done by people who are very knowledgeable about language development in deaf people; in addition, a language level should be reached that is understandable yet leaves room for growth.[18]

## THE NAD CONFERENCE

Even at this early stage, John Gough was feeling the stress of negotiating for film rights, dealing with constant delays in delivery of film from the laboratory, processing the contracts with companies, and making payments for personal services to carry out the functions of CFD. He also sensed the discontent of film companies with the slow pace of the government's operations as film orders were secured.

A month after the June 1960 conference, he stood before members of the NAD in Dallas, Texas, and told deaf participants from around the country, "In providing you with a library of captioned films, the government is giving you an equal chance to use and enjoy this medium of communication which is shared by millions of your fellow Americans."[19] But he knew this goal of an "equal chance" for deaf people to access film as a medium of communication was nowhere near being realized. The size of the program's film library remained miniscule. Gough had begun the development of CFD based on logical assumptions regarding current trends in education. Public policy regarding both general and special education was evolving with the consolidation of school districts and the expansion of their programs, post-World War II evolution of the educational system, including the GI Bill and other programs enhancing higher education, and the increased numbers of school programs for education of students with disabilities.[20] There was much to consider in the pursuit of an equal chance.

# 6
## A Terrific Start

*[W]e just do not traffic in these films at all in 16 mm so it is quite a coup to be able to offer them to your program.*

—Walt Disney Studios to Mac Norwood, quoted in James J. Kundert, "Captioned Films for the Deaf"

Despite the many challenges, CFD got off to a terrific start. An early count between October 10 and November 27, 1959, was a total of 6,075 viewers of captioned films or approximately 870 viewers per week. The total for 1960 was more than 90,000 deaf viewers of the films acquired and captioned.[1] Schools predominated with thirty out of the fifty bookings.[2]

To his credit, John Gough continued to involve deaf people in the captioned films movement, understanding the way that these films would so critically impact their own lives. After the June 1960 conference, he immediately set out to find a leader to partner with him in the CFD efforts. The person to fill the new post of program specialist, created specifically for a deaf person, would assist in the acquisition and captioning of feature films, the programming of educational films for both adult and school use, and working with adult clubs and associations to promote widespread use of captioned films among deaf people.

### MALCOM NORWOOD: PROGRAM SPECIALIST

Malcolm "Mac" Norwood was the person chosen for this important post.[3] With his experience in building a captioned filmstrip collection and a budget to rent older silent films with captions to use in classes at the West Virginia School for the Deaf, he was well prepared to begin his new adventure in expanding the use of captions on a national level. He wasted no time in beginning to persuade top-level administrators in the schools regarding the need for education of deaf children through structured use of films with captions.

In his first year as program specialist under Gough, Norwood delved into gaining a better understanding of the technical process by practicing writing titles for several films. First, he captioned the film *World Series and Football Highlights*. During the summer of 1961, he and Mary S. Larue from Gallaudet College prepared a lesson guide for the film *Rockets: How They Work*, with special attention being paid to reading and language issues.[4] Such captioned films with lesson guides that demonstrated the full capability of these films in schools across the country would reveal the economy in teaching with educational films, and encourage experimentation in their more widespread use along with other visual aids.

Norwood was the first deaf professional to join the Office of Education.[5] He played a key role in developing and refining captioning, and he also trained others (including teachers) to perform the task. He led the development of captioning guidelines, manuals, and technical specifications that were used for years. In February 1963, he began a cross-country tour to evaluate the pilot demonstration of thirteen science films, to discover their merits and observe deaf students' reactions to them.

## FILM ACQUISITIONS

Although the number of viewings in 1960 was impressive considering the limited resources, there were also many cancellations of captioned entertainment film bookings. New releases promised to CFD were held up in the captioning laboratory where titles were being etched on the films. The lab was swamped with work on 16 mm films intended for use by hearing people in South America. Brazil was an especially big user of such film technology.

Nevertheless, CFD was simultaneously building on its first efforts to develop a library of educational films. Gough invited the science department teachers at the Maryland School for the Deaf to submit twenty titles of science films appropriate for use in schools serving deaf children. Contracts were signed with Encyclopedia Britannica Films and United World Films, and negotiations were initiated with McGraw-Hill Films and Coronet Films. The selection and preparation of science films was given a priority for the pilot project. A list of thirteen science films was generated and field tested in a dozen schools, with a very positive response.

One of CFD's highlights during the early 1960s was Norwood persuading Walt Disney Studios to release prints of several celebrated productions for captioning and distribution. This included *Snow White and the Seven Dwarfs*, *Pinocchio*, *Cinderella*, *Peter Pan*, and *Lady and the Tramp* in all their Technicolor splendor. The joy of seeing these films captioned was shared by Disney's director of the 16 mm film division: "This is really quite an event for us, Mac," he wrote, "because we just do not traffic in these films at all in 16 mm so it is quite a coup to be able to offer them to your program."[6]

Mac Norwood's work created much excitement in the Deaf community. Importantly, deaf people now felt they had a direct hand in the administration of CFD. As word continued to spread around the country, rentals or purchases of 16 mm projectors increased. Entertainment films from the CFD library were shown more often in deaf clubs or private homes with groups of friends. State organizations also

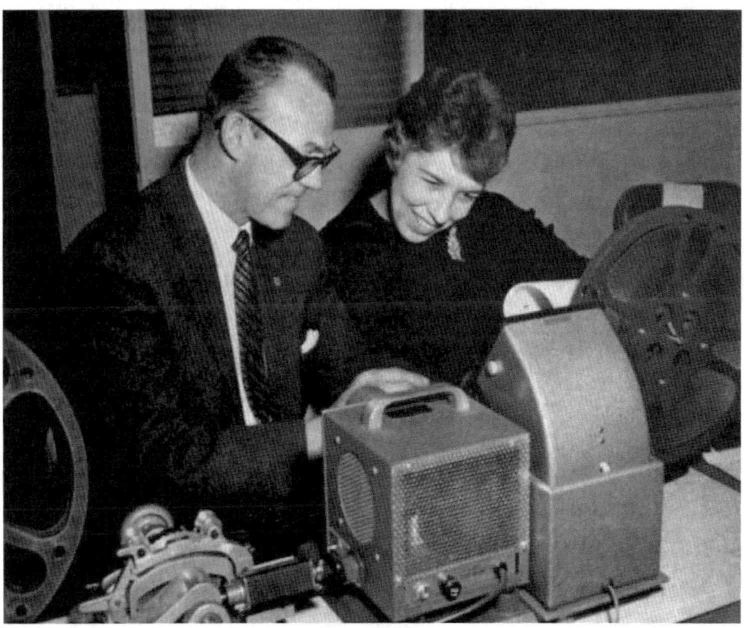

Mac Norwood and Anita Carpenter, CFD staff members, check a captioned film script against a comedy film. Image courtesy of Wallace Memorial Library at Rochester Institute of Technology.

began to coordinate their own captioning access efforts. In July 1960, for example, the United Utah Organizations of the Deaf was established to cooperate with the local organizations, such as the Utah Association of the Deaf, Utah Athletic Club of the Deaf, Salt Lake Frats, Ogden Frats, Salt Lake Valley Branch for the Deaf, Ogden Branch for the Deaf, and any other organizations serving deaf people in Utah, to coordinate dates for each group's entertainment film activities. This would reduce the chance of schedule conflicts and allow everyone the opportunity to enjoy watching the captioned films.[7]

Viewer data also showed dramatic increases. After only two years, the number of groups of deaf people CFD was serving had grown to 680, with an average viewing audience of about 18,500 people monthly. Yet, this represented less than 10 percent of the total deaf population in the United States, estimated at 250,000. To serve these people, the Office of Education acquired more than 100 captioned films. The library was soon expanded to a total of 344 prints in circulation, but this included multiple copies of some films. This growth, from an initial gift of twenty-nine prints from the American School for the Deaf, had been accomplished on a budget of $15,000 in 1960, which eventually reached the authorized maximum of $250,000 for the first time in 1962.[8]

Tabulation of film screenings for the final six months of 1962 showed that there were 1,237 film bookings with an average total monthly audience of 20,804,[9] meaning CFD had reached a total audience of 124,824 deaf people of all ages. A similar tabulation for the same period during 1960 was based on 272 film bookings that averaged thirty-four showings a month to an average total monthly audience of 3,500. Of the bookings made during those six months, 731 were made for adult organizations and 506 were made to schools and other youth groups. The wide difference in the figures was due to the simple fact that although adult organizations varied in size, the average school audience was slightly more than 176 persons, whereas the average adult audience was just under 49.[10]

> "Thank you, Mr. John Gough, and all connected with Captioned Films for the Deaf, an inspired enterprise."
>
> —*The Ohio Chronicle*, May 12, 1962

Within four more years, the total attendance at film showings would exceed 1.7 million, including thirty-four weekly programs for some 35,000 school-age children, and the funding for the program would eventually jump to $3 million. Due to the backlog of orders, however, the schools continued to have great difficulties in scheduling films. It often required booking months in advance. This itself spoke of the popularity of the captioned films.

Importantly, although few in number, entertainment films had become extremely popular among deaf people of all ages. It was an exciting time to see the promise of captioning opening up new avenues for deaf people's leisure time. In the days of silent films, older deaf people, who once regarded entertainment films as a favorite form of relaxation, now looked to captioned films like finding a long-lost friend. After one eighty-year-old man saw the underwater picture *Silent World* in 1962, for example, he said, "It was the most!"[11]

Younger school-age deaf children were also just beginning to appreciate the benefits of captioning, and CFD purposefully chose a number of film titles to cater to youth.

## DISTRIBUTION OF FILMS

Throughout the months following the June 1960 conference, John Gough gave much thought to the urgent need for visual materials for deaf learners who were challenged by language delays. Through the discussions and presentations at the conference, he realized all the more the importance of communications research to show the best ways to utilize newer media as it increasingly became available. Revision of existing materials and new research could lead to new production techniques. Gough wrote, "These and a multitude of similar questions constitute an unexplored educational wilderness which calls for the work of research pioneers to blaze new trails." The Office of Education was impressed with Gough's work in operating the CFD and the hiring of deaf staff, "The Captioned Films for the Deaf program is not only a program for the deaf, but is by the deaf as well."[12]

As Gough continued to give thought to promoting other visual aids in addition to captioned films, such as filmstrips, transparencies, and slides, he pointed out a paradox in the education of deaf students. Although this segment of the population depended almost entirely upon its eyes for learning, deaf pupils were afforded little use of *visual* aids existing in most schools. Moreover, most filmed materials at this time were relying heavily on sound for interpretation. He noted that with oral language as the chief avenue of communication, it was "erroneous to assume that sound is secondary and that pictures alone can tell a story."[13]

> "We always enjoy watching the movies shown in the chapel. . . . The title of the movie was *The Fountainhead.* We could understand the movie very well. We enjoyed it."
>
> —Ann Strickland, a deaf pupil at the South Carolina School for the Deaf and Blind in Spartanburg, *The Palmetto Leaf,* February 1961

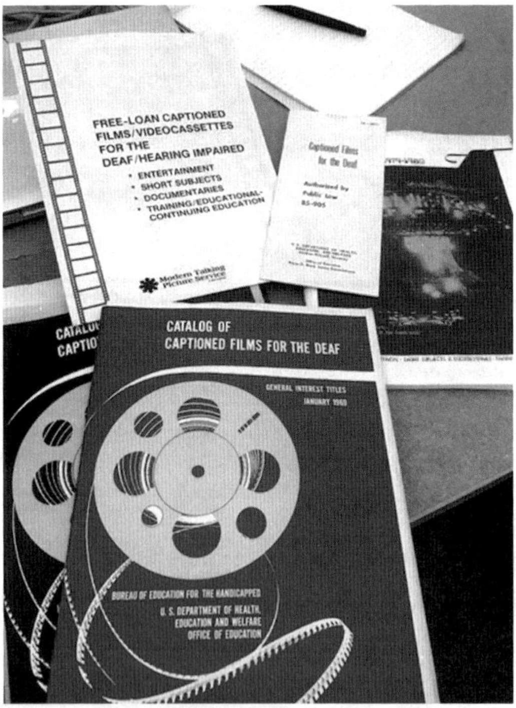


Throughout the 1960s and into the 1970s there were few captioned films in theaters and stores. CFD, the small government organization, was responsible for communicating available captioned films to schools serving deaf students and to the Deaf community. CFD media catalogs were regularly updated, published, and distributed throughout the nation.

Yet, before Gough could even begin to consider supporting the schools with visual aids, he had to solve the logistical challenge of more effectively distributing the captioned films. By April 1961, the program was serving 500 groups of deaf people registered for the captioning service. Audiences included school groups, social clubs for deaf people, civic groups, and church groups. The CFD library had expanded slowly to a total of eighty-nine entertainment and cultural prints with an additional sixty prints being captioned.

However, by September 1961, the US Department of Agriculture, which had been chosen to help CFD's efforts, had informed John Gough that it was unable to continue supporting the increased demands from the CFD program, despite being experienced with film distribution. The rapid growth of the captioning movement and limited space for the films were two of the problems. Additional facilities were desperately needed. The film library had quickly expanded to 150 films, and an average of 8,000 people were attending the showings monthly. Bookings had to be made two years in advance! Throughout the year, many Deaf clubs continued to have weekly captioned film showings. Schools were slowly subscribing to the CFD resources. Educators were just beginning to realize the potential of captioning in teacher training as well. On September 22, 1961, PL 87-276 was signed into law by President John F. Kennedy, heralding a very important step in the training of teachers of deaf students. During that academic year there were thirty-two training programs in operation either receiving federal support or having program development grants. Among the expanding priorities were the needs of media specialists.[14]

In December 1961, Gough met with deaf educators, representatives of the film industry, and the federal Office of Vocational Rehabilitation to evaluate CFD and to plan future developments. At this meeting, Mac Norwood updated the attendees on CFD's progress, explaining that in the first year the film collection reached sixty entertainment films and thirteen science films captioned on a sixth-grade reading level. There was satisfaction with the progress, but the need for more educational films had become very apparent. Also, Norwood believed that more research would enhance the effectiveness of captioned films in the classroom. He promoted the establishment of teacher training conferences, which would include demonstrations of how captioned materials could be introduced into the curriculum.

By October 1962, the Indiana School for the Deaf in Indianapolis had been selected to begin as the new national center for the collection and distribution of CFD's films. Ray G. Gallimore, a deaf teacher in the school, was assigned the role of cataloger and distributor. This assignment of the school as the national center, however, did not alleviate the problem of supply and demand. On a first-come, first-served basis, practically all available film prints continued to be booked one to two years in advance, and the booking and shipping office at the Indiana School was forced to postpone or turn down many requests for entertainment films simply because of the limited size of the library.

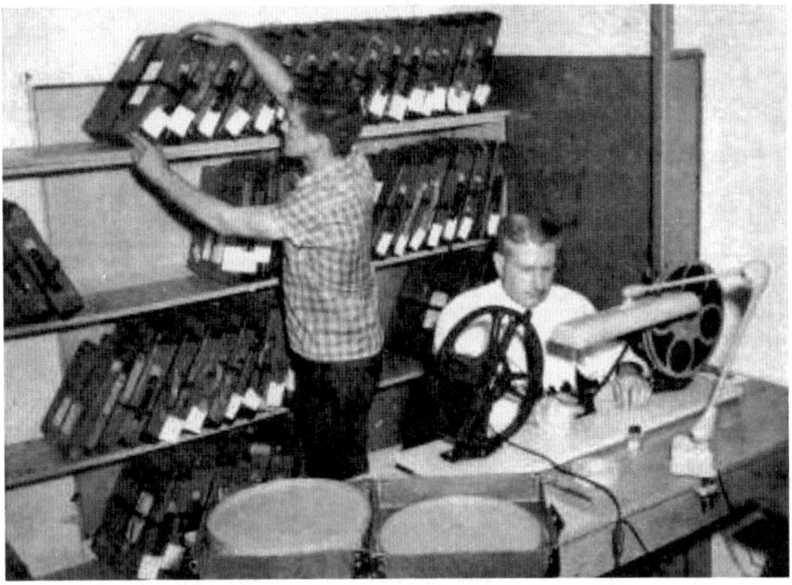

Ray H. Gallimore, a teacher at the Indiana School for the Deaf, served as director of distribution in 1961. Gallimore (seated) is shown inspecting a print that has been returned after a showing. Films were rewound, cleaned, and inspected before being sent out again. Kenneth Baughman, an Indiana School senior, is placing a print on the shelf. Image courtesy of Wallace Memorial Library at Rochester Institute of Technology.

## INTERNATIONAL CONGRESS ON THE EDUCATION OF THE DEAF

In June 1963, CFD highlighted its extraordinary accomplishments when a wide variety of captioned films were featured at the "Film Theatre" of the International Congress on the Education of the Deaf at Gallaudet College in Washington, DC. The Congress had more than 1,500 delegates registered from nearly fifty countries and from all over the United States. An entire building on campus, Kendall Hall, was set aside to accommodate the various film showings managed by CFD. In addition to about twenty films captioned for classroom instruction, the display included a premier showing of a training film especially designed to teach keypunch operation to deaf people.

As a special feature for foreign guests, a slideshow with captions in French, German, and Spanish depicted a visit to the Virginia State School for the Deaf and Blind in Hampton, Virginia. This was part of a visual aids display including filmstrips and short strips available on free loan to American schools for the deaf. The Film Theatre showed a different feature film and a selected short subject each night, including such movies as *Man of a Thousand Faces* (the story of Lon Chaney, whose parents were deaf), *Man on a String*, *The Absent-Minded Professor*, *The Last Hurrah*, and *The Baron of Arizona*.

After June 1963, Norwood ordered twenty-three new feature films. Captions were written, and these were released during the months that followed as fast as producers could get them out.[15] MGM, a new supplier for Captioned Films, provided five titles: *Somebody Up There Likes Me*; *Goodbye, Mr. Chips*; *Edison, The Man*; *A Tale of Two Cities*, and *Dr. Jekyll and Mr. Hyde*. The principal characters in two of these films were deaf, at least for a time. Rocky Graziano, whose life story is told in *Somebody Up There Likes Me*, was once a pupil at PS 47 in New York City. He later recovered his hearing. Edison was hard of hearing from boyhood and became quite deaf in later life.

As for captioned educational films, more schools were sought to help distribute them around the country, and a larger network of depositories was created. At the CEASD meeting in Riverside, California, Gough expressed appreciation to the schools for undertaking these services without remuneration. "This helps to conserve our funds for purchase of more films," Gough said, "and should rebound to the benefit of the deaf generally."[16]

The growing demand for the films continued to be a problem, and their distribution became a logistical nightmare. By June 30, 1963, there was a total audience of 310,000 admissions for captioned films. At this point, the films ranged from some of Hollywood's best to documentaries and educational films. However, as with the US Department of Agriculture earlier, the distribution also became too much for Ray Gallimore to handle at the Indiana School. By January 1964, Mac Norwood developed a new plan to divide the nation into three regions for purposes of distributing feature pictures, making it possible for teachers of deaf students to obtain captioned films on short notice.[17] The Indiana School continued as a distribution point serving fourteen Midwestern states. The eastern center, at the New York School for the Deaf at White Plains, began serving seventeen states and the District of Columbia, whereas the western center at the Colorado School for the Deaf and Blind in Colorado Springs was responsible for a territory comprised of nineteen states. This was the second major shift in distribution arrangements since the captioned film service became a government function.

As summarized in chapter 5, the need for more media in school programs was also largely influenced by the CFD program, with regional and local workshops and rapidly increasing resources for media-based instruction.

> Early in the summer of 1965, Washington, DC, was the site of the Tenth International Games of the Deaf ("Deaflympics"), the first time in history the United States served as the host nation for this event. The Office of Education placed an early bid for a film production of this historical event and CFD had the opportunity to reveal, for the first time, its ability to actually produce film materials.

In his personal memoirs, John Gough took note that his start-up years in CFD were filled with drama. His own reflections reveal how "the strenuous pace and some of the pressures involved in my work were taking their toll."[18] He wrote of one troublesome lobbyist on the staff of a congressman with substantial political clout in Washington. According to Gough, the lobbyist attempted to involve CFD in the funding of a film on the education of the deaf that would be produced under a noncompetitive contract, which Gough refused. Before long, the lobbyist was back in his office, this time brazenly requesting the authority to approve in advance all contracts with CFD. "My reply was that in no way was this going to happen."[19]

The lobbyist had also established a national advisory council but in fact, there were no deaf people on this board! Shortly after starting up, the group called for a full accounting of CFD and for Gough to be present at one of their meetings for questioning. When Gough appeared at the meeting, a man from the Bell Telephone Company challenged CFD's involvement in research, claiming that this should not be within the province of CFD but rather should be given to Bell Telephone. During this meeting, still another board member, whom Gough had known for many years, told him that the "trouble" was that he hadn't asked for any suggestions to guide CFD, although Gough had visited him in his office not long after he joined the US Office of Education and opened the door for suggestions.

These are typical examples of behind-the-scenes challenges confronted by Gough during these early years of administering CFD. They show that it was not all smooth sailing as he began to lead the program. This was especially true when the Bureau of Education for the Handicapped was established, resulting in "layers of administrative personnel" who took away the freedom he had felt during his first few years.[20]

## THE MEDIA REVOLUTION

Although the establishment of the film distribution centers had great potential for effecting positive change in classrooms, especially in the nation's residential schools, a serious problem was soon recognized that would make the efforts of CFD all the more challenging. Teachers around the country had begun "teaching captioned films" rather than using captioned films to expand and enrich curriculum offerings. Many teachers had college education in other disciplines than the primary content areas and the education of deaf students. A large number of teachers had degrees in general and special education.

Protocols were needed to monitor appropriate use of films and their relevance as tools to satisfy established curriculum requirements. Robert "Bob" R. Davila was at the time a supervising teacher and later principal at the New York School for the Deaf, and he recalled how he worked with teachers to use films in a connected way to expand and enrich their prepared lessons rather than using the captioned films as free-standing lessons.[21] Bill Stark also reflected that probably the worst use of a film or video in the early 1960s was to just fill time or entertain. "If a film was indeed strong enough to 'be' the curriculum, we would not then have been concerned about that."[22]

This problem of having deaf students view free-standing captioned films not tied to specific content being taught in the lessons was happening all over the profession, and it was a primary reason that various centers at universities began structured training to help teachers and supervisors learn how to properly incorporate films and other media when they developed curricula. It didn't resolve the issue completely, however.

With teachers experiencing the importance of demonstration or description of effective teaching strategies, a better understanding of content was also needed. CFD began planning national and regional workshops as a meaningful approach to train teachers in the most effective use of captioned films. The program had also been developing "lesson guides" for use of the films. CFD provided instruction on how to use the guides as both an introduction to a unit of study and a summary of a unit's content, as well as helpful information on how the film or video might be used to support and enrich the instructional units established by a school.

It should be remembered that many classrooms at this time had just begun to see the use of basic overhead projectors. Media such as filmstrips were also beginning to be used in schools. There was no internet available to help teachers prepare for classes. In fact, it was not until the 2000s that the program began promoting online lesson planning and tutoring of teachers on how to browse for the program's videos based on state and national standards.[23] Bill Stark led the way through the years in providing a useful resource for inexperienced as well as experienced teachers. Stark emphasized, "Students don't magically learn from a video just by staring at it. What you do before, during, and after the viewing will make video-watching a real learning experience."[24] He later offered ten rules that would not make a teacher an instant expert in the use of captioned videos, but "they could help teachers avoid some major pitfalls and get students excited about learning."[25] In the 1960s, however, there were no such resources available to teachers of deaf students. In summary, the challenge presented to CFD was to help the schools get away from the use of captioned films as pure entertainment and emphasize the instructional value of the media in the classroom.[26]

As the distribution of entertainment films slowly progressed, Gough and Norwood were pressed to proceed with the purpose of determining priorities in educational films. To garner additional funding, CFD encouraged people to write their congressional representatives and urged removal of the statutory limitation on the annual CFD expenditures ($250,000), as proposed in new Senate Bill 2511. This bill, if passed, would enable the CFD to submit its own annual budget requests.

As a result of hard work on the part of CFD, educators were beginning to see the immense potential in the revolution taking place in the use of media in the classrooms. However, such efforts lacked coordination nationally. CFD staff were sensing the rapidly increasing need among teachers who were contacting the program and suggesting films that had potential educational value. Robert Baughman, for example, principal of the Kentucky School for the Deaf vocational department, wrote to CFD to recommend captioning a short film on hockey. The film had been shown without captions at his school following a full-length motion picture. Baughman pointed out that with captions, the film would have been an excellent lesson in Canadian geography as well as on the vocabulary associated with hockey.[27]

Other types of media and the methods of using them were also gaining attention. Importantly, the Ohio School was experimenting with the use of visual aids in a lighted room, which would allow teachers to simultaneously communicate in sign language.[28] Benjamin Schowe Jr., a teacher and librarian at the school, had furthered his education through courses in visual materials at Ohio State University. At the school in Columbus, he developed an impressive program of motion pictures, filmstrips, and other visual aids.[29]

With Ohio in range of airborne television broadcasts from Purdue University in West Lafayette, Indiana, Schowe also investigated how the Ohio school could exploit this medium. Interested in sharing what he was learning, he became the columnist in the *Ohio Chronicle* for a column titled "Visual Aids and Library Notes"—a digest for educators in other schools serving deaf children to inform on what was happening in the visual education field.[30]

However, unable to make effective use of some audiovisual materials provided for hearing children, schools for deaf children remained in dire need of captioned resources. The seriousness of this situation was highlighted in a 1962 national survey of visual education in American schools for deaf children conducted by Gallaudet College. The report stated that about half of the schools for deaf children and more than half of the day classes had no facilities for producing their own visual aids. "Even where such facilities exist, they tend to be rudimentary. At the same time, the most frequent comment by the administrators who feel that their schools do not make sufficient use of visual aids is that their teachers lack the necessary training."[31] Forty-two percent of the schools and 36 percent of the day classes "made no provision for in-service training."[32]

Importantly, CFD was cited as the most frequent supplier of visual aid materials. It was also pointed out that the 200 or more day classes for deaf children made little use of these films whatsoever; it appeared that residential schools were taking advantage of the program much more.

Although the investigator conducting this survey concluded that "The need for more publicity about this service among the classes is evident," Gough and Norwood believed that the real need was probably not for more publicity but for more materials of an educational nature, which would be of greater use in both schools and day classes. "By expanding the deaf child's experiences through the use of films," Gough stated, "and by helping him to verbalize these experiences through printed captions, modern technology can bridge part of the communication gulf which separates the deaf from an adequate education."[33]

## STRESSING LITERACY THROUGH CAPTIONING

Norwood and Gough traveled to various cities to monitor the distribution of media and to describe in detail CFD's plans to expand educational and recreational film availability. When Norwood met with the Utah Association for the Deaf in Ogden, where there was a sizable and appreciative audience served by CFD, he answered many questions from the audience and exhibited several short films. As in other locales, the Deaf community there hoped that the captioned films might help lead educators to a significant breakthrough in education, especially in terms of literacy. Deaf people were deeply indebted to Norwood for taking the time to show what was in store for them. A former teacher himself, Norwood capably emphasized that the greatest challenge a deaf child has in learning is the acquisition of language and that deaf children have a better chance now through captioned films to associate language with what they see. The seminal PL 85-905 (1958) was a great step forward for the education of deaf children because it provided the opportunities for developing and distributing good materials with a strong emphasis on visual learning.

Gough also continued to emphasize the value of captioning for language development, explaining that the young deaf child "faces one of the most insuperable of all educational barriers—that of learning language without any basic language on which to build and without the natural means of acquiring this basic tool, namely, hearing."[34]

## PUBLIC LAW 87-715: PRODUCTION AND DISTRIBUTION OF EDUCATIONAL AND TRAINING FILMS

On August 7, 1962, John Gough summarized the results of the Gallaudet survey before a Special Subcommittee of the Senate Committee on Labor and Public Welfare, which had been tasked with deliberating Senate Bill 2511, a bill to expand the CFD program. In his testimony, he also drew comparisons with the Books for the Blind program, revealing a much higher 1962 budget for that program—approximately $1,800,000 in serving an estimated total blind population of 350,000—as compared with the current $250,000 budget for captioned films. In addition, book publishers were waiving royalty rights on books used for the talking books program whereas full-length films normally carried a royalty of $600 per print for a five-year lease in the captioned films program. Even more upsetting was that educational films costing

$60 per print included a royalty item, which sometimes ran 75 percent of the purchase price.[35]

Educators believed that passage of Senate Bill 2511 would do for visual aids in teaching deaf students what educational television was doing for the use of visual media in teaching hearing children. Importantly, preparing for this bill brought further realization that education had been overlooking the immense potential of visual aids for teaching deaf children. It was a critical turning point in history and CFD took a lead role in the revolution.

Elliott Newcomb, a representative from Encyclopedia Britannica Films, who was also one of the participants at the 1960 conference, felt it was incredible that the acceleration in the public school program resulting from the Russian-launched Sputnik had side-stepped the education of deaf children. Determined to do something about it, Newcomb spoke two weeks later to the Senate. He was a former campaign manager for Senator Edmund Muskie (D-ME), who, although familiar with the education of deaf children in his state, needed Newcomb's encouragement to pursue the issue further.

Originally sponsored by Senators Muskie and Claiborne Pell (D-RI) and by Representative John Fogarty (D-RI), Senate Bill 2511 stressed the increased use of visual instructional materials on a national scale. Muskie, described as, "six-feet four, powerful voice, big head, very imposing, and could be an intimidating figure," had been in the Senate only four years, but he was already an important participant in national politics, developing a reputation "as an expert in writing and enacting legislation."[36] During the special subcommittee hearing concerning this bill, Muskie showed his strong support for aiding the education of tens of thousands of deaf and hard of hearing children. "The more I have studied the potential of new methods, particularly audiovisual aids," he said, "the more I have become convinced of the tremendous possibilities of this method in teaching the deaf."[37] He emphasized that the time had arrived to take positive action to assist deaf students.

Muskie's advocacy on September 7, 1962, garnered the support of educators, school administrators, and producers of educational films. Maurice B. Mitchell, president of Encyclopedia Britannica Films, Inc., noted, "The logic of the motion picture in the classroom is so apparent as to almost make it unnecessary to labor the point."[38] But government support was needed to bring this crucial accessibility to deaf students because film producers were not willing to make any investment of capital in the venture.[39]

In his testimony, Gough summarized that the importance of enlarging CFD was amply clear. This was not only to provide more films but also to explore new techniques for teaching deaf students. Gough

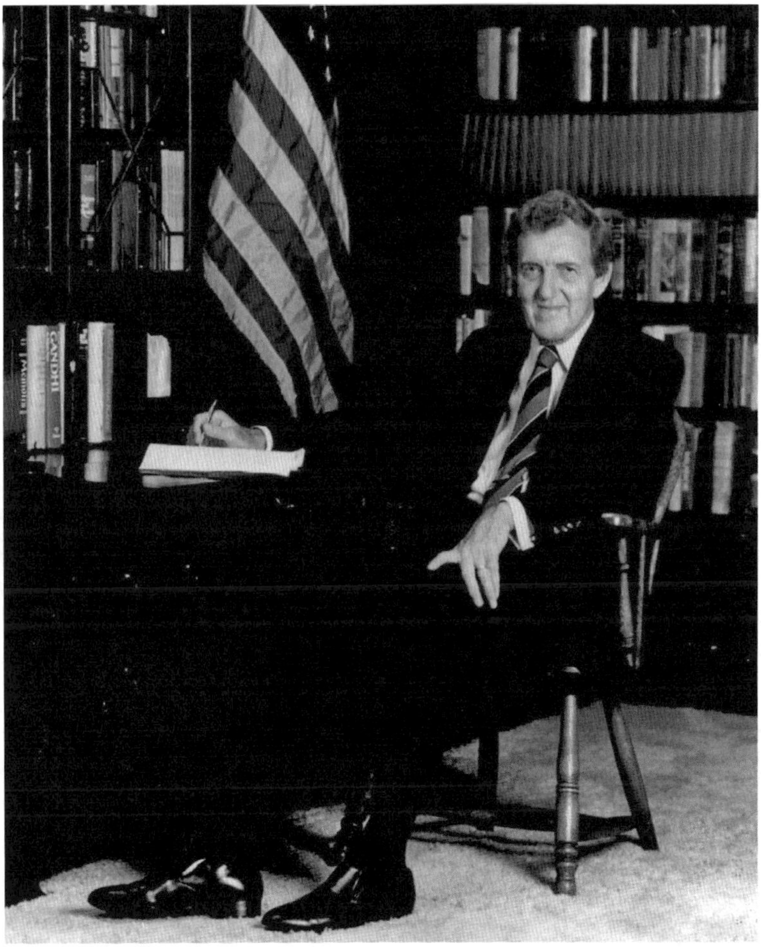

Senator Edmund S. Muskie was an influential advocate of captioned educational media for deaf and hard of hearing students in the 1960s. Image courtesy of Library of Congress.

also emphasized the importance of increased training of staff at schools serving deaf students in order for them to make better use of the films.[40]

Senator Pell similarly expressed his strong support: "The proper education of the deaf is a national challenge which deserves our serious attention. It is clearly important that we provide our deaf children with the best possible learning tools and that we help to stimulate adult education programs for the deaf which will in turn enable them to enhance their contribution to our society."[41]

Despite CFD's success, more activism among the Deaf community was needed to move the pending legislation. The NAD encouraged its members to make their opinions known. The Maryland Association of the Deaf formed a campaign to bring the bill to a favorable vote. George Pratt from the Clarke School for the Deaf in Massachusetts generated support, pressing for members of the Committee on Legislation of the Council on the Education of the Deaf to join the other groups and individuals. In addition, Patria Winalski of HEW's Congressional Liaison Office, who had a deaf son, exercised considerable influence among her colleagues.

With strong backing from constituents and legislators, the bill eventually passed without a single dissenting vote in the House or the Senate, becoming PL 87-715 on September 28, 1962. Significantly, it amended the 1958 CFD authorization by adding a mandate to "promote the educational advancement of deaf persons" as well as to carry on research, training, production, acquisition, captioning, and distribution of educational and training films. John Gough's program had a remarkable track record over the first year. His innovative vision for CFD was realized through this legislation, giving the program greatly expanded authority in the areas of materials development, training, and research, along with significantly increased appropriations. With the enactment of PL 87-715, captioned films services broadly expanded, with the annual appropriation increasing from $250,000 to $1,500,000. CFD was now serving more than 700 schools and organizations for deaf people in the United States.[42]

The same year, the Eighty-Seventh Congress passed a second piece of significant legislation affecting the lives of deaf people. Signed by President John F. Kennedy, PL 87-276 authorized a two-year accelerated program for training teachers of deaf students. In addition, the

National Institutes of Health began supporting numerous research projects related to deafness. Added to this, the Office of Vocational Rehabilitation was advancing studies and planning the development of facilities for deaf adults.

With the passage of PL 87-715, teachers were elated with the additional funds that would enhance the education of deaf children. Deaf adults saw the legislation as also providing exciting new leisure time experiences and cultural awareness for deaf people of all ages. At the American School for the Deaf in Hartford, captioning pioneers Edmund B. Boatner and his associates took special pride in how they had fostered an idea into a million-dollar operation. The American School celebrated the passage of PL 87-715, and Mac Norwood, a graduate of the school and former teacher there, was invited to be a guest speaker. In gratitude, Norwood brought a flag that had flown over the Capitol during the Eighty-Seventh Congress for presentation during the ceremony. Attending were Congressman John Fogarty (D-RI), who had championed the films bill in the House, and Patria Winalski, who had successfully steered the bill clear to passage.

# 7
## The Ball State Workshops

The advent of CFD and the magnitude of its impact on the deaf population of the United States resulted in quick recognition of educational media as a tremendous but relatively untapped educational resource. The federal legislation passed in 1962 also called for a new emphasis on research and an accelerated program for teacher training. This taxed CFD to the point that there was much concern about staffing.

### CFD EXPANDS STAFF

For the production of filmstrips, Gough also hired Richard Simpson, a twenty-year Navy man with experience in photography and film direction, as the CFD production specialist. Edward C. Carney, a deaf man formerly with the Jewish Rehabilitation Service in St. Louis, Missouri, was recruited as the materials specialist. As with the earlier appointment of Mac Norwood, the Deaf community looked to his appointment as a show of solidarity.[1] Carney was also appointed reports officer and public relations specialist. Gough then hired still another deaf man, James J. Kundert, as assistant program specialist to supervise the preparation and processing of caption scripts for educational films. This position was also created due to increased demand for the CFD's services. Kundert became an assistant to Mac Norwood.

The extension of the benefits of media to all individuals with disabilities led to the Division of Media Services in the Bureau of Education for the Handicapped, with two branches—the Captioned Films and Telecommunications branch and the Learning Resources branch. As they worked together, these two branches attempted to set up a systematic approach for enabling children with disabilities to acquire the skills they need through a comprehensive network of media centers providing a wide variety of services.[2] John Gough and Mac Norwood were very much behind the new thrust on instructional media that led the program to go far beyond the distribution of captioned films. In an article titled "Captioned Films for the Deaf: The New Program," Gough described the evolution of CFD. Testimony in favor of the new PL 87-715 had hammered at the educational and language lags of deaf students and the promise of captioned films to help improve their education. With the budget ceiling increased to $1.5 million, training activities were now authorized as well as the production of filmed materials for use in educational TV, filmed textbooks, and other means of improving the teaching of deaf students in the country.[3]

One barrier CFD faced at this time, however, was that in their rush to adjourn, Congress failed to pass a supplementary bill that would have given CFD the funds to move ahead. Another barrier was one of commitment. Gough knew very well that the future success of CFD in contributing to the education of deaf students would depend on the reactions of deaf people themselves as well as school administrators and teachers.

The US Office of Education recognized the importance of continuing the pursuit of access through captioning, including entertainment films, but it also saw the networking created by Gough and Norwood in CFD to be a useful prototype for enhancing academic success in deaf students. At this time in history, most schools for deaf students were just beginning to take advantage of recent advances in material production technologies. Many were expanding their library and resource staff to include individuals "recently trained in current and emerging media technology production and utilization techniques. These included the use of heat, light, and dry ink particle technologies to produce permanent paper copies, transparencies, and laminated materials."[4]

## GILBERT L. DELGADO HIRED TO COORDINATE WORKSHOPS

In response to the need for an experienced educator of deaf students with administrative skills, John Gough selected Gilbert "Gil" L. Delgado from the California School for the Deaf to coordinate the teacher training that was being planned for CFD.[5] They had known each other from their interactions at various deaf education conferences. Delgado was impressed with Gough's enthusiasm about the ambitious project, but it was at first difficult for him to leave his family in California to tackle the logistics of the summer workshops. This would include developing criteria, curriculum, and processes for selecting participants;

planning work spaces; lodging; rental of equipment; and shipping large amounts of media. Although challenged by limited funds to implement this project and pay all of the staff and participants, Delgado nevertheless felt honored by Gough's need for his creative talents. "Gough was not someone you could say no to!"[6]

Before CFD could accommodate the teachers who were clamoring for instructional films, it was first necessary to find a way to coordinate curriculum development across the country—if that were possible. There was no desire to create a national curriculum, but the hope was to find the most effective way to integrate captioned films with particular lessons and, as stated earlier, to help teachers avoid using films as standalone lessons. CFD had experimented with teacher workshops on a small scale and was receiving compliments about the success. One teacher from El Cerrito, California, for example, wrote, "I sing your praises at every opportunity. I can't show you the end results, but the vocabulary growth [of the deaf students] can be heard every day and that is music to my ears."[7] The schools also encouraged CFD to consider an extension of the filmstrip service,

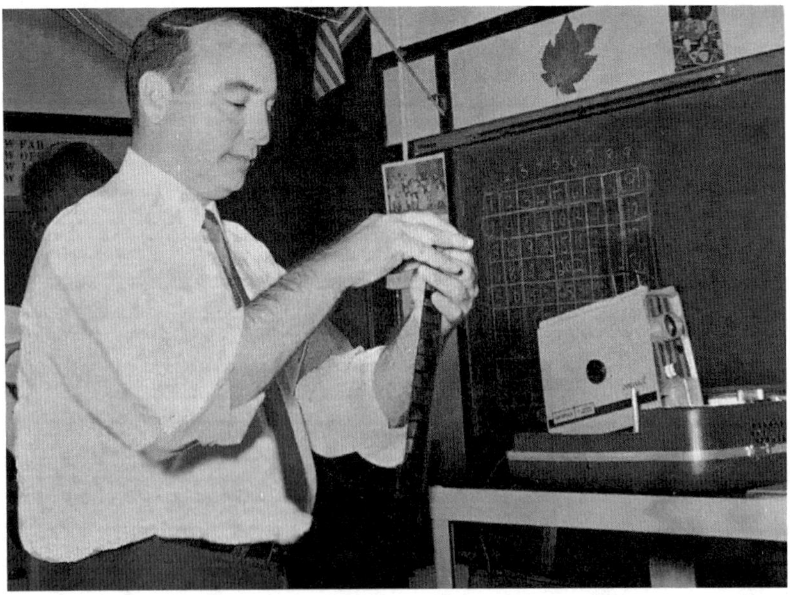

Gil Delgado with a film strip during the early days of the revolution in the use of media in the education of deaf and hard of hearing students. Image courtesy of Gallaudet University Archives.

developing a series with captions appropriate to the vocabulary level of deaf children in grades 1–6. Every indication was that teachers in the field were more than ready to learn how to best use the captioned films and other visual aids. John Gough seized the opportunity at this time to put into effect the contracts CFD had drawn up earlier with a number of film producers. The overarching goal was to determine common factors and to provide guidelines for CFD's selection of visual aid materials to caption and distribute to schools. The underlying belief was that expanding the curricular offerings of schools and programs with improved visual aid support would accompany improved quality of content and presentation skills of the teachers. This could be achieved through the participation of teachers in a series of intensive interactive workshops. Many teachers in the schools at this time in history had some training in educating deaf students, but they lacked formal training in the content areas. And with the media being new to them, they were not familiar with how best to incorporate visual aids in their instruction. The workshops Gough envisioned were patterned somewhat after the existing National Defense Education Act Workshops that were underway around the country. He saw a seminal project providing the seed to spawn numerous future efforts to enhance curricular offerings to deaf and hard of hearing students. These ambitious plans awaited the hiring of additional staff.

Delgado outlined the series of curriculum workshops for teachers of deaf students based on the priorities established by a national steering committee after a systematic assessment of needs. In June 1963, the Office of Education awarded a grant to Ball State Teachers College (later, in 1965, Ball State University) to host CFD's first summer workshop. Because the workshop was held at Gough's alma mater, he was familiar with personnel and facilities at this institution, and he could move quickly to get the project implemented. The college had already established an extensive library of curriculum materials for schools serving deaf children. It also offered an excellent general library of educational resources. The "Ball State Workshop for Improving Instruction of the Deaf" emerged from many discussions with educators around the country.

The workshops centered on several curricular areas. In the summer of 1963 the topic was social studies and literature enrichment. In 1964 the area of concentration was on science. In 1965, an omnibus workshop covered the related fields of personal-social, sex, and consumer education. In 1966, mathematics was the topic,[8] and in 1967 it was vocational education. This design for improving curricula in the schools serving deaf students was in harmony with national curriculum movements. There was an emphasis on having the teachers attending the workshops evaluate as many contemporary visual aids as time would allow in the six-week sessions, especially films and other materials specific to the curriculum area chosen for each summer. As an outcome of the workshop sessions, the visual materials most directly applicable to teaching deaf learners would be recommended to CFD for distribution. Workshop staff were experienced teachers who created a lesson guide for each captioned film, which included the learning goals for the product, ideas for teacher and student preparation for its use, and follow-up activities. The guideline manuals from each summer's workshop were printed, bound, and shipped to teachers across the country.

Each summer a group of twenty-five to thirty experienced teachers was selected based on demonstrated teaching skills and knowledge in particular subject areas. The participants convened for a six-week period of intense activity, which included receiving instruction in use of associated media and applying that knowledge and skill to the production of the curricular guideline manuals. These manuals utilized textbooks and other print materials currently in use by the schools. They emphasized suggestions for use of the visual media. To complete this task, workshop participants were given access to a wide array of commercial media and materials. As was done with the film *Rockets: How They Work*, individual lesson guides were written for each captioned educational film to ensure their effective use. During the workshops the teachers also identified additional films to be captioned by CFD for use with the curriculum materials.

The first twenty teachers from state and public schools and colleges arrived in Muncie, Indiana on July 18, 1963 for Phase I training. Phase II (second year) of the Curriculum Workshop had a research focus this time on applying contemporary methodology to curriculum development and instruction.

Delgado served as the coordinator of the workshops in Phase I and II. He was supported by several Ball State School of Education staff. Alan Huckleberry was his liaison, and Louis E. Inglehart, a journalism professor, was also very helpful. The master teachers who were workshop participants developed the curriculum guides at various grade levels around

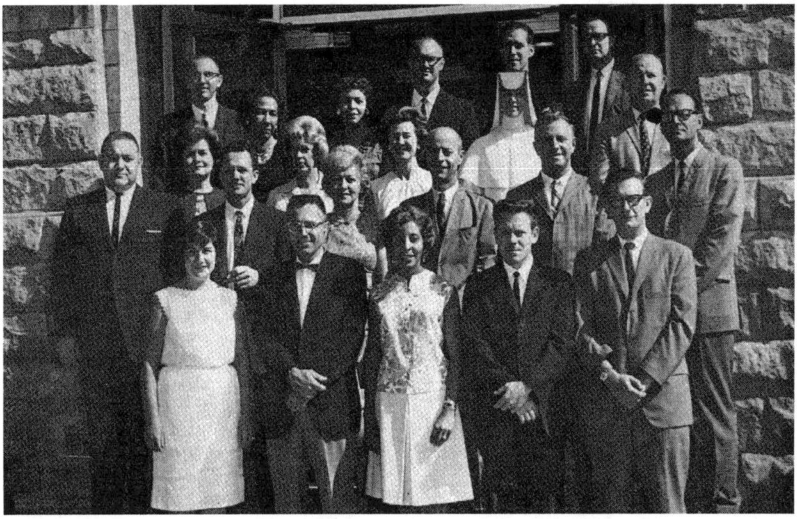

Workshop at Ball State University. Captioned Films for the Deaf was sponsor of a five-week workshop held at Ball State University in the summer of 1965. This workshop focused on personal, social, and economic education. Participants included, first row, left to right: Despena E. Varkados, Ben M. Schowe, Jr., Roberta Malone, Harry J. Murphy, Jr., Jerome W. Freeman; Second Row: Doin Hicks, Charles Overholser, Caroline Burnes, Gregory W. Jones, Leonard F. Becker, Emil S. Ladner; Third Row: Mary Jane Harris, Margret F. Gillespie, Carleda S. Moore, Sister Gilmay Rudicus, Roy K. Holcomb; Fourth row: Robert T. Dawson, Elaine Johnson, Rubye Stamper Frye, Paul Peterson, Gary A. Curtis, Kenneth W. Kritz. Image courtesy of Wallace Memorial Library at Rochester Institute of Technology.

instructional films to aid in classroom instruction.[9] Powrie V. Doctor from Gallaudet College served as a language consultant and Mac Norwood was the expert on technical aspects of the captioned films.[10]

Gough placed complete confidence in Delgado to proceed with the new curriculum and instructional strategies. One decision was to plan one of the summer workshops to focus on vocational education. Boyce Williams saw great potential in vocational education and rehabilitation of deaf people, and was concerned about the slow progress made for this population as compared to that achieved for hearing citizens. The services and resources were seldom available to deaf people with comparable effectiveness "because of their communication problem, as technological advancement has closed employment doors of traditional values to the deaf. . . ."[11] Between 1945 and 1960, a total of 19,136 deaf people had received vocational rehabilitation services. In 1964 alone, that number exceeded 2,300. Thus, Williams welcomed the efforts to further increase awareness of vocational rehabilitation resources among those attending the workshops.

Delgado hired Williams as a consultant. The deaf expert had entered the vocational rehabilitation program about eighteen years earlier. His responsibility was to develop a rehabilitation program for deaf people that would serve to compensate in part for the severe limitations in opportunity for economic adjustment attending the disability of deafness. He was amazed at the barrenness of what was available in 1945 and the slowness of the rehabilitation program's progress.[12]

In addition to vocational technical high schools, schools for deaf students as well as some general high schools continued to provide a significant amount of vocational training. Williams was only too glad to help Delgado develop the workshop devoted primarily to

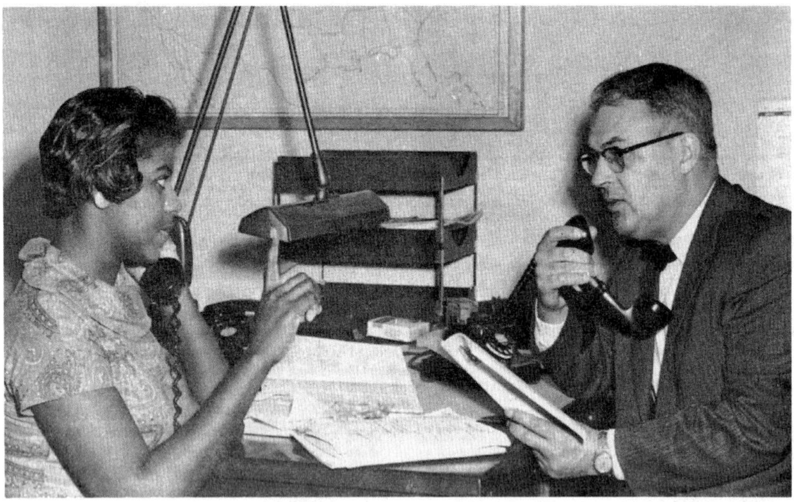

Boyce Williams, a deaf consultant from the Office of Vocational Rehabilitation in the Department of Health, Education, and Welfare (HEW), welcomed the opportunities to increase awareness of vocational rehabilitation resources among those attending the Ball State workshops. Charlotte Coffield is shown interpreting a telephone call for him. Image courtesy of Gallaudet University Archives.

pre-vocational and vocational curricula with the new emphases on captioned visual materials.

Another major concern challenging Delgado was that of locating and hiring a person to supervise the selection and acquisition of the large amounts of media materials that were needed for workshop participants to review. Ben Schowe Jr. from the Ohio School and Alfred J. Lamb, assistant superintendent of the Indiana School, planned the workshops for 1964. Once the media were acquired, the task of making them readily available to participants would also need to be addressed. Schowe had extensive experience in curriculum design and the use of visual media. An exemplary teacher with superior organizational skills, he was key to the success of the workshops.

The first week of each summer workshop focused on orientation and instructional activities with emphasis on visual aids. Participants in the first workshop represented a cross-section of schools and programs.[13] Teams of two or three worked on specific courses or areas of curricular content. A professional editor assisted the teams with any concerns they might have in developing their materials. Each team expected to have a finished product ready for final editing and reproduction by the end of the six-week workshop.

The curriculum guide manuals developed from the workshops were concentrated on courses most frequently offered in middle school and lower division high school levels. They included content outlines, suggestions for activities and skill sets that students should be able to demonstrate. The manuals were primarily for teacher use and were not intended to replace texts and other materials provided by school systems.

## DELGADO MOVES TO THE NATIONAL CFD HEADQUARTERS

Delgado's effective organization of the Ball State project and administration of the first two summer workshops was instrumental in his being invited to join Gough at the CFD headquarters in Washington, DC, as administrator of the newly expanded media initiatives. On September 15, 1964, Gil Delgado entered his role as a member of the program staff in Washington. In his new role he continued as an oversight officer responsible for the media revolution taking place.

He also filled the long-vacant position of Research and Training Specialist.

Doin E. Hicks, a participant in the Phase I workshop, was selected to replace Delgado as the on-site summer workshop coordinator. His professional career had begun as a classroom teacher at the Missouri School for the Deaf and this was followed by four years as principal at the Arkansas School for the Deaf.[14] Hicks oversaw the remaining three summer workshops. Overall, the five workshops achieved the goal of producing a total of thirty-nine curriculum guideline manuals. Importantly, the master teachers who were participants received a rich understanding of the value of incorporating instructional media, including captioned educational films. About 150 teachers participated in the summer workshops and most became good ambassadors for encouraging use of the materials among colleagues upon returning to their home schools.

After the workshops were completed, the staff continued to promote use of the curriculum materials, especially the visual aids. Hicks's experi-

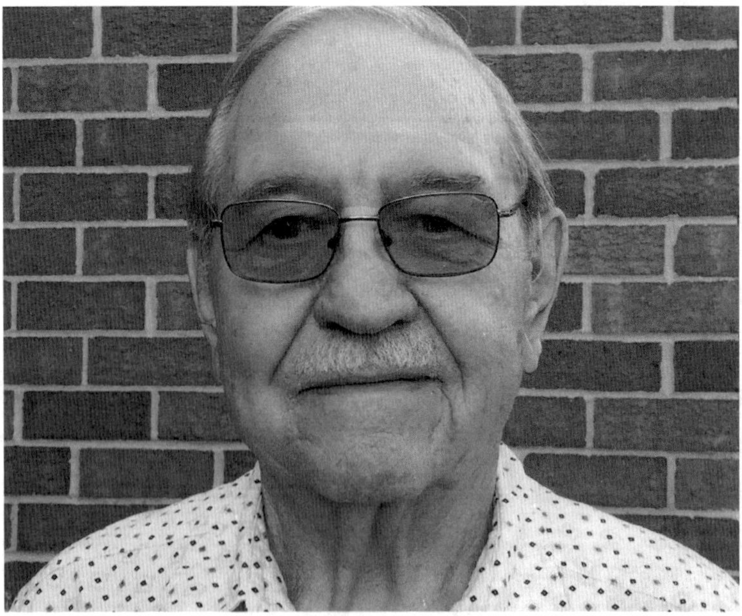

Doin Hicks, with experience as a classroom teacher at the Missouri School for the Deaf and as principal at the Arkansas School for the Deaf, oversaw the final three workshops at Ball State. Image courtesy of Doin Hicks.

ences at the workshops, for example, left a deep and positive impression on him, and for the next thirty-five years continued to influence his career as an administrator in schools and programs and as an educator and consultant preparing teachers and administrators to serve within the profession. Following the Ball State experience, CFD funded a proposal Hicks submitted on behalf of Callier Center in Dallas, Texas and the Pilot School Division. The proposal was for support of a project to develop an individualized instruction system for deaf students with heavy emphasis on the use of visual media as proposed within the Ball State guide recommendations. As a component of the project, a multimedia delivery system was devised, including a special desk which put controls for film, filmstrip, and overhead projectors at each teachers' fingertips. The project had other useful components as well, including research demonstration, parent participation, and dissemination of information about effective instruction.[15]

One lasting product of the Ball State workshops was the increased awareness of the value of captioned educational films in the classrooms. The schools depended greatly on CFD's annual catalogs of available films. The workshops and other early demonstration projects clearly indicated the leadership role played by CFD in its quest to open new doors to deaf and hard of hearing students in their pursuit of learning.

# 8
## The Captioning Program Expands

*We are excited every time we have a captioned film. We like them better than those we really do not understand because they are talkies.*

—Barbara Smith, *The Deaf Oklahoman*

While offering the Ball State workshops in 1964, the staff of CFD also inaugurated a demonstration of a series of thirty-one filmed lessons for the training of IBM keypunch operators. These lessons were originally developed by the Perceptual Development Laboratories of St. Louis for use with their projector, known as the "Perceptoscope." CFD added captions and clues for deaf trainees so that, hopefully, the films secured by the Office of Education would be better understood with minimal help from instructors. These filmed lessons were ready to use by early March 1964.[1] They were tried out at demonstration sites in six locations: Riverside City College in California, Utah School for the Deaf, Florida School for the Deaf, Pennsylvania School for the Deaf, Virginia State School, and the Division of Vocational Rehabilitation in the District of Columbia. Lloyd Ambrosen, president of the Convention of American Instructors of the Deaf and superintendent of the Maryland School for the Deaf in Frederick, was director of the field project. Deaf and hearing instructors from each of these locations were brought to Washington for two weeks of intensive training.[2] CFD's staff believed this offered potential not only for specialized training programs such as the keypunch but also for the teaching of reading and other skills. CFD was interested in the development of captioned electronics courses, including mathematics and other content. John Gough's program had not only provided increased access through captioning, but it was contributing to adult education through captioned vocational training films. When the keypunch workshops were completed, the instructors left for their respective centers to conduct a three-month course for deaf trainees. This was particularly notable because in the wake of automation, career opportunities for deaf people were diminishing.

Filmstrips and transparency sets produced at CFD quickly became popular, particularly the visual language transparency sets created

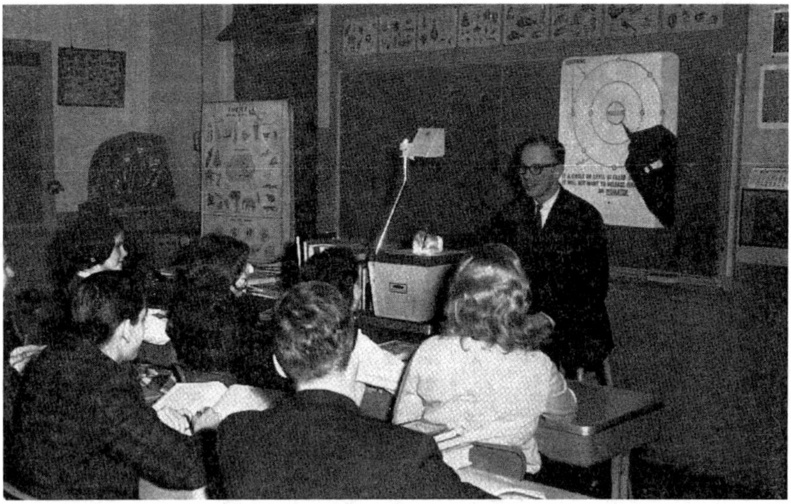

At the American School for the Deaf, science instructor Frank Ashlar uses an overhead projector to explain the structure of an atom. *The Deaf American*, October 1965, p. 3. Image courtesy of Gallaudet University Archives.

and produced by Robert F. Newby, media director at the Pennsylvania School for the Deaf.[3] In 1964, a new captioned educational film, "Human Growth," had its first screening at the NAD convention in Washington, DC. This was the first sex education film captioned for deaf students; it was received favorably and was released to schools that fall.[4] That year CFD's nationwide distribution system included the fifty-eight captioned educational film depositories located in residential schools throughout the country and an evaluation and selection system to determine which films, educational or theatrical, were added each year.[5] Counting educational films, filmstrips, short strips and other low-cost items, the library inventory was estimated to number more than 30,000 items by June 1964.

Networking across the nation was paying off. Educators of deaf students, and deaf people themselves, in other countries, especially those with English language users, were also sending inquiries to CFD staff about the possibilities of extending its benefits and services. Requests came in from Canada, England, New Zealand, and Jamaica. Even non-English countries like Germany and the Netherlands contacted CFD.

When the Illinois School for the Deaf celebrated its selection as a depository by CFD, Bill Stark, who worked as the depository manager, carried out the typical responsibilities Ray Gallimore had in Indiana—checking out films, reporting on audiences, and promoting CFD. As with many other depository managers, Bill was fortunate to have cataloging and other library skills. Without such knowledge, a depository might fail, affecting the national program. Captioners had a less glamorous responsibility than depository managers, but managers like Bill were the backbone of CFD and helped significantly in making it succeed. Stark was destined to eventually take the directing role in the national captioned films program.

One problem was that films were returned to the school depositories without being rewound on the reels, and it was a time-consuming task to prepare the films for the next users. Some schools hired deaf students to help with that. Before a film was shipped to another location, an inspection was also needed for cleaning and repairs. This often required splicing broken films, and care had to be taken to reduce the jumping effect resulting from such repairs.

At first, a room in the Illinois school's administration building called the "Audiovisual Aids Room" housed only twenty-eight captioned films. In contrast to the media we offer deaf students today, this collection of films seems minuscule, but the excitement of finally having captioned media especially focused on the needs of the deaf learner was genuine.

> "One Saturday night we went to the auditorium to see a captioned film about 'The Story of Louis Pasteur.' We really enjoyed it. It was good that we could learn so much from it."
>
> —Barbara Smith, the Oklahoma School for the Deaf in *The Deaf Oklahoman* 53, no. 5 (March 1961).

## EVALUATING FILMS FOR CAPTIONING AND DISTRIBUTION

Around the same time that the Ball State workshops were begun, CFD conducted other workshops to screen, select, and recommend films for captioning on a national level. The process of evaluating, adapting, and distributing the captioned educational products began early each fall when curriculum requests were prioritized for purchase and sent to thousands of producers and distributors of educational media across the country. Distributors were invited to submit titles that matched the specified curriculum needs in the school programs. Some distributors sent CFD new releases in all curricular areas.

At fifteen field evaluation sites, groups of educators of deaf students completed formal, written evaluations of about 500 preselected film titles. Each evaluation team included at least one deaf educator. This ensured that the "deaf perspective" was accurately represented for each title. A sign language interpreter was used to convey the audio content of the film or video to the deaf evaluator(s) in the viewing groups.

Most of the criteria by which a film title was judged were similar to those considered by any educational materials evaluation team: educational value, technical quality, viewer interest, accuracy, flexibility of use, bias or stereotyping, and organization, among others. There were, however, additional criteria specific to a deaf audience. For example, evaluators working for CFD were expected to check to see if a filmmaker had conveyed the film's message *visually* as well as aurally. Deaf people were much more dependent on their vision than hearing people; so films that used little motion, or that failed to correlate the visuals and audio in a meaningful way, were thought to be less successful with deaf audiences.

Some films and videos that were highly successful with hearing audiences often did not adapt well for deaf audiences because of specific production techniques. Therefore, educators of deaf students who were caption script writers also critiqued each film on the basis of "captionability." Overuse of flowery or technical language, too many short scenes, montage effects, rapid-fire dialogue or narration, audio lead-ins (i.e., a new scene begins aurally before it begins visually), and visuals that do not correspond directly to the audio were production techniques that created captioning hazards.[6]

Final screening of each film and video occurred at a national workshop. The participants included master teachers in deaf education, administrators, captioning experts, experienced caption and lesson guide writers, and media specialists. About two-thirds of the evaluators were deaf. Both selection and validation teams evaluated such aspects as a film's basic educational value, content accuracy and suitability, potential viewer interest, potential use, and captionability.[7] The validation groups included experienced caption and lesson guide writers and consisted of teachers, administrators, media specialists, and other professionals in the field of deaf education. A list of selected film titles was then recommended to the Department of Education for purchase and adaptation.

On March 12–14, 1964, the Office of Education called a conference of school administrators, educators, audiovisual specialists, and rehabilitation workers to evaluate CFD's developments to date and to suggest directions the program should take. The conference participants were impressed with the progress that had been made with educational films in a few short years.

As for captioned theatrical (entertainment) films, in 1965 Gough and Delgado expanded the staff and coordinated the national distribution through three special depositories in White Plains, New York (eastern region), Indianapolis (central region), and Colorado Springs

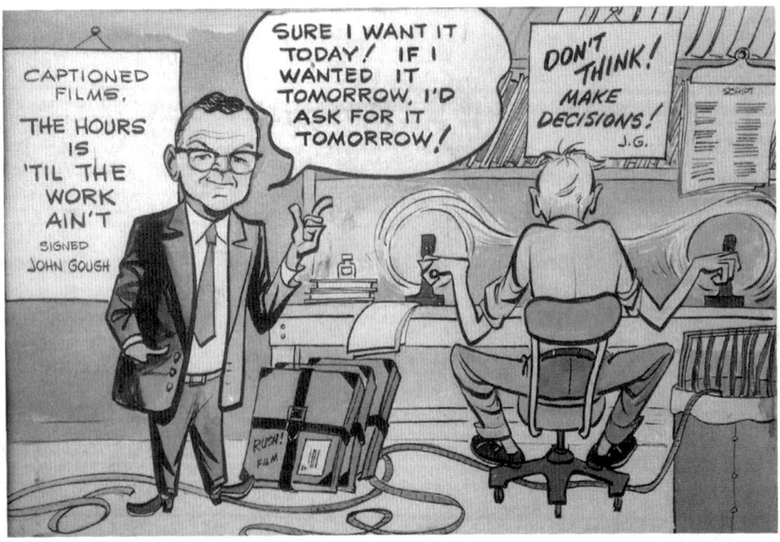


John A. Gough was a hard-nosed leader, dedicated to bringing access to deaf people through captions. At one point, he was teased for his straightforwardness through a cartoon (artist unknown). Image courtesy of Susan Gough Updegraff.

(western region). Each center was responsible for eighteen to twenty schools.[8] At this time, also, the NAD was contracted with the Division of Handicapped Youth, Department of Health, Education, and Welfare to evaluate films of general entertainment. The NAD was also charged to create criteria that would be used to determine the suitability of available films for captioning. Frederick C. Schreiber, treasurer of the NAD, was in direct charge at this time. Approximately 100 deaf people and a few hearing people took part in evaluating the entertainment films.[9] Within a few years an estimated 200,000–250,000 deaf children and adults were viewing captioned entertainment films. CFD had certainly come a long way in its first four years as a federally funded program.

John Gough and Mac Norwood were busy one day when an unexpected visitor called on the CFD headquarters in Washington, DC. Gough's daughter, Susan Gough Updegraff, recalled the incident. It was during the height of the Cold War, and an FBI agent showed up at the office and asked for Gough. The agent asked Gough what business CFD had with iron curtain countries. Gough answered, "None," and the FBI agent told him that one of the CFD staff had been seen entering the Russian embassy. The FBI had cameras focused on the embassy.

Gough called in a staff member who explained to the agent that he had learned about a Russian film having to do with deaf people and he had gone to the embassy to see what he could learn about possible use of the film in teaching deaf students.

"So the contact was explained," Susan wrote, "and no one was accused of treason!!! I think CFD never got a hold of that film but it made for a good story."

—Susan Gough Updegraff, email to author, December 22, 2018

As time went on, CFD continued to add new media projects, including the development of color slides to reinforce idiom concepts, color films for use in primary and elementary grades, training films for use in modern mathematics instruction, 8 mm cartridge loop films for fingerspelling and speechreading practice, films for graphic arts instruction, and instructional materials for a typewriting course.

Other workshops also were contributing to the sizable additions to CFD's educational film library. For example, a caption writing workshop held at the Minnesota School for the Deaf during the summer of 1965 focused on new films, many of which were recommended by a curriculum committee that convened at Ball State. Another workshop held concurrently at the California School for the Deaf in Berkeley devoted itself to the preparation of specially designed lesson guides to accompany captioned educational films that had been developed at the Ball State workshops. These guides were devised to enable teachers of deaf children to utilize the films more effectively and to enable deaf children to absorb more of the concepts therein. As such, they would supersede commercial guides prepared by many of the larger educational film suppliers expressly for use in regular classroom. As the media revolution continued, many more workshops were provided around the country at individual schools.

A variety of captioning projects was also being led by CFD. One project Delgado helped to develop was a new system that promised to make captions much more readable. Working under a government contract, a private film company in Washington had developed a new machine that placed a uniform shadow behind all captions. This made the captions stand out much more clearly and promised a definite improvement in the readability. Inserting a dark background behind captions made them more visible when appearing with such images as arctic snow, jungles, woodlands, or grassy settings that often made the captions illegible.

Anticipated research included fifteen in-service short course demonstrations by the New Mexico Foundation, three six-week media institutes for thirty teachers each, special saturation projects at the Arizona and Arkansas schools, and symposium on a systems approach in the instruction of the deaf at the University of Nebraska. There were also production projects over which Gough and Delgado had oversight, including captioned filmstrips for *My Weekly Reader*, which was based on the *New Science Adventure* series, and *My Summer Weekly Reader* for first- and second-grade readers. Nearly 12,000 deaf and hard of hearing students on the first- and second-grade reading levels subscribed to *My Summer Weekly Reader*. CFD believed introducing this service would enable deaf youngsters to continue learning while at home during the summer months.

While these projects continued, the letters from film viewers and their families kept pouring in, showing great satisfaction with the media and other support Gough's program was providing. "It is wonderful to know that our son is cared about," wrote the parents of one deaf boy from Du Bois, Pennsylvania, to John Gough, "and we thank you for giving Stephen the opportunity to have the films and *My Weekly Reader* free."[10] Two parents from Warwick, Rhode Island, wrote, "This is truly an exciting way for a deaf child to continue his education during the summer months. I am afraid [this note] cannot adequately express my surprise and appreciation of these most useful materials."[11] Another educator from Miami, Florida, wrote, "The Thomas Jefferson Junior High School class for the Deaf would like to thank you for playing such an important part in their general and cultural advancement. The educational and training films were invaluable. Our interest in your program has been further stimulated. We look forward to continued use of the program. . . . "[12]

To accommodate the growing demand for captioned films, Gough set up still another caption writing workshop at the Austine School for the Deaf in Brattleboro, Vermont with six trained teachers devoting four weeks to the preparation of scripts; and a second workshop at DePaul University in Chicago to focus on the accompanying study guides. Teachers responded well to these materials. For example, Margaret McCormick, Principal of the Willis and Elizabeth Martin Public School in Philadelphia, emphasized in a letter the worth of the workshop products. "The guide to be used with captioned films for the deaf is excellent. In fact, it is the most useful aid for meaningful teaching of the deaf to come to my attention in recent years. . . . I should like to take this opportunity also to tell you how very valuable are the captioned films received at Martin."[13]

To ensure that teachers would have access to the media without waiting, the filmstrips were shipped directly to the schools to be kept on a free-loan basis and returned only upon request.

> "I would never have understood why the movie was so titled without the captions."
>
> —Sharon Kay Wood in her school newspaper, *Arizona Cactus,* December 1961, after watching *God Is My Co-Pilot*

## CFD'S RESEARCH AND DEVELOPMENT PROGRAMS

The mid-1960s included reports trickling down to CFD staff that many on Capitol Hill recognized Gough's substantial achievement. These hinted that efforts of Congress to ratify legislation permitting an extension of CFD's activities looked promising.

During their tenure together, the roles of Gough and Delgado were often intertwined, and their accomplishments were impressive. They were visionary leaders, as was Norwood, initiating innovative and critically needed educational technology and telecommunications. They coordinated the captioning of entertainment and educational films at as rapid a rate as budgets would allow. They also coordinated a workshop on film distribution and developed CFD's loan service regulations.[14] They were both educators, and they knew that even a good educational film could lose much of its effectiveness if it was not used properly.

As Gough contemplated how CFD could address the innovative use of media in classrooms serving deaf students, he envisioned several principal activities. The US Office of Education had begun designating certain major activities under a "special project" heading. There was Project Sights and Sounds, Project LIFE, Project Guidepost, and others. Such use of media apparently gave these research activities added credence. Project Sights and Sounds, contracted with the American School for the Deaf, was focused on developing a set of six filmstrips to accompany 33 1/3 records and picture test booklets for enhancing use of residual hearing in primary-age children. The high quality of the work led by Professor Harold Saleh of Connecticut State College in New Haven earned him the American Film Festival's 1965 blue ribbon award for the best filmstrip in the arts.[15]

Project LIFE (Language Improvement to Facilitate Education) was a very ambitious project in the development and testing of elementary language materials for young deaf children. It assumed primary responsibility through a new approach to language instruction using a filmstrip program. Gough turned to the National Education Association to carry out the objectives he had developed. Harley Z. Wooden, former Michigan School for the Deaf superintendent, was project director, and the distinguished Arthur I. Gates of Teachers College, Columbia University, was involved with one of the trials. This was a many-phased nationwide study of the use of multimedia aids for developing English language skills in deaf children.

Two things were important in order for Project LIFE to mature. Qualified people would need to stay with the project, and teachers in the classrooms would need to critically evaluate the materials that came from the project. Because commercial producers of educational material are seldom familiar with issues unique to deaf children, CFD encouraged teachers of deaf children to share their ideas for developing the media.

Project Guidepost focused on the development of instructional materials for use in vocational guidance. Yet other projects focused on strengthening the visual perception of deaf children and developing the Demonstration Project in the Use of Educational Television for Educating and Informing the Deaf and Hearing-Impaired. In addition, there were efforts, some of which CFD carried out through contracts with other agencies, to enhance library services in school programs serving deaf children, typing training for deaf people, and the teaching of speech. A project designed to teach rhythmic speech to deaf children and one for the use of audiovisual material in acoustic training for vocabulary and language building were also conducted.

Delgado provided able supervision of the research and training activities that fell under Gough. With his personal experience as a Hispanic scholar and former classroom instructor, Gil also initiated a project titled Springboards to Language, which attempted to provide an additional tool for use in language classes that might help deaf children express themselves with more proficiency in English. This project included a set of slides used to enhance creative expression. Teachers found other uses for the slides, including vocabulary building, teaching adjectives, and crafting open-ended stories.

The problem Springboards to Language encountered after the Ball State workshops were completed was that the rich supply of captioned films and other visual aids generated from the Ball State workshops and the many research and development projects were being underutilized. In general, teachers serving deaf students were not familiar with the new materials. Importantly, they were bewildered by the new captioning and other media technology being brought into their classrooms. Only about 150 teachers had been formally trained during the five summer workshops.

Attempting to address this problem, a contract with the Office of Education, the New Mexico Foundation, Inc., was created at New Mexico State University. This demonstration project was directed by Marshall Hester, retired superintendent of the New Mexico School for the Deaf. In this new training venture, Hester emphasized the heavy use of visual media. He visited many schools and programs over a wide area to disseminate information about the use of captioned films and other visuals.

Based on Hester's success, Gough immediately began to formulate the establishment of other centers at universities providing him and his staff remarkable information dissemination support, including regular publication of successful techniques and strategies in the *American Annals of the Deaf*, the major deaf education journal at the time.

## PUBLIC LAW 89-258

In a letter to his predecessor, Edmund B. Boatner, John Gough expressed the excitement of the day. On June 18, 1965, he wrote, "It looks as if there will be a new bill introduced soon to increase our program. I would scarcely have believed it two years ago but our present appropriation is already slim for running a program of the dimensions that this one has grown to be. Your baby has grown to be a pretty lusty youngster."[16]

Then, on June 30, 1965, Senators Pell and Muskie introduced Senate Bill 2232 to the Senate Committee on Labor and Public Welfare. The bill emphasized the use of all educational media, rather than films alone, and for the distribution of such media to parents of deaf children and others who are directly involved in advancing the learning of deaf people or who are actual or potential employers of deaf people. PL 89-258, passed by the Senate on August 31, by the House on October 5, and signed by President Johnson on October 19, 1965, increased funding for CFD from $1.5 million to $3 million for years 1966–1967; to $5 million for years 1968–1969; and to $7 million thereafter. The increase in funding allowed CFD to purchase and distribute much more media equipment to school programs. It was a welcomed boost.

## MEDIA SERVICES AND CAPTIONED FILMS

The law also led to the consolidation of CFD into a new federal office renamed Media Services and Captioned Films (MSCF) to reflect the broader focus on media. MSCF became a branch within the recently formed Bureau of Education for the Handicapped. Services were also extended to individuals with other disabilities. In effect, the Media Services part of the branch comprised everything except captioned films. The Captioned Films part maintained the roles CFD had played, fulfilling the original intent of that program to caption and distribute movies to adults and educational and entertainment films to schools.[17] Despite this name change, the program name "CFD" persisted in use for years.

One objective of PL 89-258 was to promote the general welfare of deaf people by "bringing to such persons understanding and appreciation of those films which play such an important part in the general and cultural advancement of hearing persons" and "providing through these films, enriched educational and cultural experiences through which deaf persons can be brought into better touch with the realities of their environment." Other objectives included promoting the educational advancement of deaf people by "carrying on research in the use of educational media for the deaf," "producing and distributing educational media for the deaf and for parents of deaf children and other persons who are directly involved in work for the advancement of the deaf or who are actual or potential employers of the deaf," and "training persons in the use of educational media for the instruction of the deaf."[18]

As acquisitions officer for MSCF, Norwood had spent millions of dollars in procuring films, filmstrips, slides, and transparencies for deaf audiences. Interestingly, despite the increasing emphasis Gough and Norwood had sensed in the government stressing educational films more than entertainment films, the legislation continued to promote "cultural advancement" and "cultural experiences" for deaf people. Both were associated more with entertainment films.

By 1966, the acquisition of prints and related educational materials generated during the Ball State workshops expanded CFD's media library by 356 percent, bringing the total to 203,127 filmed items, 5,135 prints of seventy-nine educational titles, and 576 prints of forty-eight general interest titles. In addition, there were 1,020 prints of fifty-one free or sponsored subjects, 216 educational titles, and 216 general interest titles totaling 24,249 motion pictures.[19] The remainder of the CFD library consisted of filmstrips, kinescopes (film recordings of television broadcasts), slides, transparencies, and related materials.

With the nearly fourfold increase in CFD's library, there were 1,372 groups registered for film service in 1966 in comparison with 1,100 at the close of fiscal year 1965.[20]

## REGIONAL MEDIA CENTERS

An important element of the federal government's emphasis on expanded educational media in the schools serving deaf students was the establishment of regional media centers for the deaf. The idea for these centers first came up at the time legislation was being prepared for CFD in the late 1950s. C. Walter Stone, in the New Educational Media branch, Department of Health, Education, and Welfare, Office of Education, had asked John A. Gough to serve as interim director of his branch while he took some time off for a project in Africa. Under Title VII of the National Defense Education Act, research proposals were pouring in to be ranked and rated by an advisory board. Among these was a modest request to produce films to teach young deaf people how to read lips. It was submitted by Robert E. Stepp Jr. at the University of Nebraska. The proposal was funded and, in time, through meetings between Gough and Stepp, the concept of media centers to help promote the captioned films agenda emerged.[21]

During this same time, the passage of the Elementary and Secondary Education Act (ESEA) of 1965 made possible further nationwide growth in the availability of public school media equipment and materials. Taking advantage of ESEA, nearly every classroom in the country's residential schools was equipped with much-needed video programming devices. As these schools acquired the projection equipment, filmstrips, and other software, demand for the program's captioned educational films also grew.

In addition, the operation of the four regional media centers established in the summer of 1966 helped MSCF address the priorities of dramatically increased research, especially in the area of language development, earlier identification of deafness, and increased parent education.[22] Each of the four regional media centers had a major emphasis. The Southwest Regional Media Center at New Mexico State University in Las Cruces focused on curriculum with an emphasis on programmed learning and instructional systems. Promotion of visual education in the Southwest was carried out by Marshall S. Hester, who led the original demonstration project at New Mexico State. These centers were invaluable to the media revolution taking place in school programs serving deaf students in the United States.

The Southern Regional Media Center at the University of Tennessee, under William D. Jackson, focused its attention on educational television. This center served a twelve-state region with the overall purpose of conducting training, research, and development in the area of educational media for the improvement of instruction and services. It also hosted an annual six-week media institute during the summer.

The Northeastern Regional Media Center for the Deaf at the University of Massachusetts in Amherst, under Raymond Wyman, emphasized distributing media through a clearinghouse, preparing media demonstrations, and conducting a summer media institute for teachers of deaf students. The major undertaking of their project was to establish a transparency center. Transparencies were developed for a variety of specific teaching tasks. A file of more than 10,000 masters was assembled.

The Midwest Regional Media Center for the Deaf at the University of Nebraska, under Robert E. Stepp Jr., had as its first function the goal of designing, developing, adapting, and producing instructional materials for the express purpose of teaching deaf students. This center also conducted research into the effectiveness of these materials in an area field testing program, and planned and conducted the first Symposium on Research and Utilization of Educational Media for Teaching the Deaf and an educational media institute for teachers of the deaf.

All four regional media centers adapted and produced various visual media, including speechreading and speech-training films, filmstrips, programmed learning, videotapes, pictures, multimedia packages, and transparencies. The centers also supplied overhead projectors, filmstrip projectors, screens, and projection carts to residential and day school classrooms and training centers. More than two million discrete items of media and instructional software were made available to teachers during this project.[23] Depository managers were among the attendees at the training workshops and symposia conducted by the centers.

Through the four regional media centers, MSCF continued its extensive efforts to influence improvements in the instruction of deaf and hard of hearing students. A specific example of this was a series of summer teacher training workshops held at the Southwest Center in New Mexico. At that point, Donald Ferguson, who had supported the workshops at Ball State, moved to New Mexico State as associate dean of the School of Education and assumed administrative responsibility for the regional media center located there. He assisted with many of the training activities and continued throughout his career to be a valuable spokesperson and supporter of efforts to improve educational opportunities for deaf students.

## EARLY EFFORTS TO DEVELOP EDUCATIONAL TELEVISION

Gough's program also oversaw developments in educational television using closed circuit, educational television, and open broadcasts. More than 700 organizations were endorsing MSCF, some in a formal manner and others by their use of captioned educational films. Things were changing rapidly in the Department of Education, and Gough saw the implications for CFD very early in his tenure. PL 87-715, passed in 1962, had emphasized education over entertainment. A few months later, in March 1963, Pittsburgh was chosen by the US Office of Education to develop a set of ten half-hour training films for teachers of deaf students. WQED, the educational television station in Pittsburgh, was given $53,666 in federal funds to produce a series of videotapes. Cooperating in the project was the Western Pennsylvania School for the Deaf in Edgewood, where classes were taped in half-hour segments and used in the University of Pittsburgh graduate courses for preparing teachers.[24] The films illustrated utilization of visual aid materials in instructional scenarios. The project was successful and the films were distributed to both pre-service and in-service training programs.

Other instructional television experiments included productions of speech lessons at the John Tracy Clinic in Los Angeles, and at the Arkansas School for the Deaf in Little Rock there were pioneering attempts to present regularly scheduled television programs for deaf people through local commercial stations. However, although general education had sought and was receiving massive funding from government and foundations for television facilities, equipment, programming, and experimentation, and had developed large numbers of qualified people, special education had not.

Through Gough's program, the Department of Education had made an excellent start on the rapidly growing media revolution and, especially, bringing access to spoken communication to deaf and hard of hearing people through the process of captioning. In 1963, Norbert Nathanson was working as a producer/director of closed circuit credit courses (learning by television) for the Education Department of New York University (NYU) at its Washington Square Campus in New York City. Nathanson's degree project consisted of a comparative evaluation of two matched groups of deaf students from the New York School for the Deaf to be taught identical subject material by a teacher at the school. The first group was taught by the traditional face-to-face methodology, and the second group was taught by specially prepared television programs produced and viewed live in a classroom at NYU. Both groups were then tested for learning growth, and the results were compared. John Gough asked if he could attend one of the televised programs that Nathanson had developed. In the discussion that followed, Gough learned of Nathanson's interest and expertise in television technology, particularly as it related to teaching. Gough invited Nathanson to participate in the evaluation process used by Captioned Films for the Deaf to fund projects of developing interest and, subsequently, to attend a meeting at the Midwest Regional Media Center for the Deaf and speak about television experimentation and potential use in the education of deaf students. Nathanson's perspectives on instructional television would benefit CFD over the years to come.

By 1965, however, interest in using television for the education of deaf students was virtually nonexistent. In the summer of 1966, Gil Delgado called Norb Nathanson, told him that John Gough had decided to create a national television network for deaf students, and asked if he would be interested in coming to Washington, DC, to head up the venture. Nathanson was delighted and accepted the offer, but unfortunately, after he had moved to Washington, President Lyndon Baines Johnson froze all new hiring. Gough could do little but facilitate a one-year contract, hoping that the needed funds would be approved by the end of the year. Nathanson nevertheless began studying all of the activities of MSCF and meeting the personnel nationally involved in MSCF in order to explore how his talents and training could be of value as he became familiar with the program. In this initial pursuit, Nathanson spent many days traveling all over the nation with Mac Norwood, studying sign language, meeting deaf and hearing associates, and becoming familiar with the MSCF program, especially the potential of captioning.

However, as Nathanson's first year came to a close, Gough came to him with the disappointing news that funding would not be forthcoming in the next year. Nathanson continued to consult with MSCF as needed. When the Southern Regional Media Center for the Deaf presented a videotape production at the MSCF-sponsored Lincoln Symposium and shared a set of guidelines, for example, the center was immediately bombarded with requests for information and guidelines. Nathanson had become a leading advocate of instructional television for people with disabilities and was an asset to MSCF. Yet, although general education had explored, experimented, adapted, and employed television to good purposes and with good results, the same could not be said for special education.[25] There was no organized or systematic approach to television research, no readily available body of information on the subject of television for people with disabilities, relatively little source material, no widespread professional activity in this area of activity, no compendium of historical facts, and no data on costs or effectiveness.

Thus, a lot more work remained to be done in relation to the use of instructional media with deaf learners, especially in the area of educational television. Among the expanded services provided to teachers of deaf students resulting from PL 89-258 of 1965 was the establishment of the Educational Media Distribution Center in Washington, DC, the following year. This center distributed films and disseminated services. Building on the instructional media revolution that had been taking place, MSCF supplied 4,000 classrooms and training centers with basic equipment consisting of overhead projectors, filmstrip projectors, and tables. Included in the free distribution program were the curriculum guideline manuals.

## PICKING UP MOMENTUM

In 1965, 16 mm films distributed through the depositories were being captioned at the rate of about sixty-five titles per year. The responsibilities of the MSCF program, however, involved much more than the captioning and distribution process. A mechanism for screening and selecting films for captions was established, and lesson guides were developed annually for new films entered into the educational library. The NAD agreed to provide support for the most time-consuming activity—the evaluation and selection of Hollywood productions. Conveniently headquartered in Silver Spring, Maryland, a suburb of Washington, DC, the NAD recruited a cross-section of deaf people, including housewives, college professors, students, teachers, printers, and clerk typists, to screen uncaptioned films submitted by companies. The individual reports were consolidated into a single report to Gough's office. The drawing of evaluators from the professional organizations serving deaf people possibly increased the chance that a common cause, captioning for all deaf people, might increase the chance that the captioning program might encourage opposing oralist and signing camps to work more closely together.[26]

More workshops were also conducted for writing captions. This sometimes involved rewriting the entire soundtrack. Data collected indicated that schools, organizations, clubs, and eligible agencies exceeded 19,000 film viewers per month. The educational films were seen by more than 1.5 million deaf and hard of hearing viewers every year.[27]

Gough's program was circulating 125 educational films and 212 feature films, and although the focus was shifting toward more educational films than feature films, some of the educational titles used in the schools were also being made available to deaf adults.

On November 9–10, 1965, MSCF sponsored a two-day workshop called "Applications of Systems Approach to the Education of the Deaf" at the Arizona State School for the Deaf. The meeting was the first of a series to be held in preparation for a symposium on the use of visual aids and similar materials in the teaching of deaf pupils to be held at the University of Nebraska the following spring. Robert E. Stepp Jr. was the chairman of the workshop. Attendees included leading deaf educators such as Ray Jones (San Fernando State College); Leo Conner (Lexington School for the Deaf, New York City); William D. Jackson (Pilot School for the Deaf, Dallas, Texas); and Marshall Hester and John Gough of MSCF.[28]

The Midwest Center at the University of Nebraska stressed the utilization and production of 8 mm films. Of great success and renown was a one-week annual educational technology symposium for administrators, teachers, media specialists, and other school media personnel. Gough described the University of Nebraska as a "place with a lot of potential," adding that the officials there also expressed much interest in national symposia. Stepp had done some unique and interesting research on using 8 mm films. He would be developing classroom systems to allow the equipment to be easily manipulated. Four schools in the Lincoln, Nebraska, area were selected as demonstration centers for several years "to reveal what can be done, trade information, and discover problems of residential and day schools" with regard to the use of media.[29]

MSCF had a noticeable expansion of the national library of captioned films during all of these research and training activities. Between 1965 and 1966, the number of educational motion picture titles had increased from about 200 to more than 325. The number of general interest motion picture titles during this same period increased from about 260 to more than 300. Free or sponsored films produced by federal agencies and commercial concerns provided vocational and/or general information as well as instruction in such areas as driver education and medical self-help. Ford Motor Company provided the driver education films, which were captioned by MSCF. Union Pacific and Santa Fe Railroad companies provided travelogues, which were helpful in social studies classes. Prevocational courses used films donated by the American Federation of Labor Committee for Industrial Organizations. The National Aeronautics and Space Administration was the source of films on missile industry and space program information. In the field of elementary science instruction, MSCF negotiated for the acquisition of part of the celebrated *Mr. Wizard* television series.

Across the country, deaf people continued to gather together to see the films. They loved the movies as well as the social occasion they offered. As Jamie Berke recalled, "We saw the films in a community college classroom. People brought their own snacks—popcorn, candy, or even real food. After the movie was over, everyone socialized, especially the families with children. All the movies that were shown were films suitable for all ages."[30]

> Nancy Abbott was about ten years old when the movie *Man of a Thousand Faces* was released. She recalled, "This film was especially significant because the main character, Lon Chaney, had deaf parents, which made it important to members of the deaf community. It was why it was captioned and shown at deaf clubs across America. This film was a hotly discussed film at the Akron Club of the Deaf (OH), and I remember seeing it with my parents and their Deaf friends there. At this time, there were no objections to hearing actors playing deaf roles. Just the fact that 'deaf' people were part of the film's story was important. My parents and their friends were very excited about seeing this film with captions."
>
> —Nancy Abbott, email to author, December 22, 2018

Both the Department of Health, Education, and Welfare and the MSCF received enthusiastic letters from those using the materials. Special to John Gough, Mac Norwood, and Gil Delgado were letters about how the deaf children in the schools were enjoying the captioned films. Some of the younger boys at the Florida school, for example, had taken their scout canteens filled with ice water to the film showing and were well prepared to enjoy the evening. They watched the captioned film

*All Mine to Give*, a story of immigrants who came to America to make their home.[31]

It was at this time that John Gough realized that the consolidation of the educational media distribution contracts into a single contract with a single agency would enable MSCF to proceed more readily. The CEASD was selected for this purpose. The organization was composed of 110 heads of schools for deaf children and approximately 125 associate members representing teacher training centers, speech and hearing centers, audiology clinics and other agencies, and it was capable of reaching more than 80 percent of all teachers and nearly 100 percent of the schools serving deaf children. In June 1966, official approval was given for the distribution contract with CEASD. CEASD, under contract with MSCF, operated the Captioned Films for the Deaf Distribution Center, which also had direct responsibility for writing captions and lesson guides for the films.

By March 1967, nearly 11,000 requests for films were filled. Requests continued to flood MSCF, but unfortunately, budget restraints did not allow all of the requests to be filled.

> "Another wonderful teaching aid for our deaf children, re: your new filmstrips!. . . . We're especially thrilled with the idea of corresponding the filmstrips with the *Weekly Reader* since all of us at St. Francis De Paula use this periodical with all of our pupils. Now they are twice as valuable."
>
> —James J. Kundert, "Captioned Films for the Deaf" (unpublished manuscript). Courtesy of Susan Gough Updegraff.

## THE 1967 NEBRASKA SYMPOSIUM

George Propp, a deaf educator, believed that there had been more change in the education of deaf students during the two or three years since MSCF began its workshops than in the previous 100 years. "To date," he wrote in the *Missouri Record* in December 1967, "this statement hasn't provoked a serious rebuttal. The changes of which I speak go deeper than the conspicuous hardware and related materials, and they are reflected most profoundly in a rising note of optimism and a more positive attitude on the part of both teachers and students."[32]

Propp knew well how much technology had impacted the lives of deaf people. He enjoyed talking about it. In 1967, at the Third Symposium on Research and Utilization of Educational Media for Teaching the Deaf at the University of Nebraska in Lincoln, he gave an example of how much captioning meant to him. He described how he had obtained his elementary education in a one-room Nebraska school and was, in his own self-assessment, "technologically deprived" during his formative years. The first arc-projected 16 mm educational film he ever saw was around 1930. It was a film developed by the Department of Agriculture for county agents to show the use of electricity on the farm.

Propp's personal introduction to technology as a deaf person occurred some years later and, interestingly enough, concerned radio. Harvey T. Christian, a math teacher at the Nebraska School for the Deaf, was one of those people who applied available technology to the unique communication needs of deaf students. In the late 1930s, he rigged up some ropes and pulleys on a chalkboard to communicate World Series radio broadcasts to deaf students at the Nebraska school. By manipulating a drawstring, he was able to provide very realistic play-by-play reporting of the games, thus bringing radio communication and baseball vocabulary to the deaf students. "If this wasn't communication technology," wrote Propp, "it is only because we lack agreement on the meaning of the term."[33]

At the third Nebraska symposium in 1967, a number of papers were presented by experts from outside the field of education of deaf students. They suggested ways in which the "new" technology in education might be applied to the field. This included lipreading lessons on 8 mm films developed by Frank B. Withrow and a multimedia package in language instruction. Plans for the acquisition and distribution of equipment (e.g., overhead and filmstrip projectors) were also discussed. A second purpose of this symposium was to try to reach school administrators with a significant message about the media revolution. The coordinators considered the symposium "a most stimulating and useful experience to come together under these circumstances."[34]

John Gough also gave a short greeting at the meeting, but due to a budget crisis he had to depart almost immediately after to Washington, DC. Up until the 1967 session of Congress, MSCF had been in what was termed the Salaries and Expenses budget of the Office of Education. This was the money that was allocated by Congress to operate the office, pay salaries and consultants, and for related things such as travel and conference costs. With the tremendous growth of the Office of Education, however, and the number of demands that were made by President Johnson for various kinds of conferences and programs, there was a shortage of about a million dollars. Gough's program had about $700,000 left for the last quarter of its operations, and he rushed home to defend that budget. Sadly, he was not completely successful, and there was a very substantial reduction. For months, the government was able to fund existing projects that had been established but was not able to fund new initiatives.[35]

Despite these pressures, Gough continued to emphasize the priority he had established to involve deaf people in the program's activities. "Something was expressed yesterday in one of the papers," he stated in his remarks at the symposium, "to the effect that deaf people don't want to think for themselves, they don't want to learn, and so forth. It seemed rather negative in its estimate of deaf people." Gough told the audience that George Propp had proven himself an excellent leader, presenter, and educator and "would recognize in a moment that he is a deaf person who does not conform to that kind of an estimate." To his credit, he added, "I could point out many similar deaf people."[36] He also went on to summarize that MSCF's film acquisition program was handled on the national level by a deaf man, Mac Norwood. "He does an outstanding job. I mention this . . . to emphasize the fact that there is no stereotype of the deaf and that we need to look at them as individuals."[37]

In February 1968, at the Fourth Nebraska Symposium on Research and Utilization of Educational Media, MSCF described how the work of the centers had proceeded well during the budget shortage, especially at the Midwest Center in Lincoln under the leadership of Stepp. The theme in 1968 was "Designing Instructional Facilities for Educating the Deaf Student: The Learning Module." During the past year, the Midwest Center had been putting most of its production effort into the 8 mm films. The most significant contribution of the Midwest Center at the school level, however, was the contributions of its Media Institute graduates who entered various educational programs throughout the country. The Nebraska symposia served well to keep educators updated with the most recent breakthroughs in research and practice.[38]

The role of the captioned films program had evolved in the early 1960s to one that included many forms of visual media as well as teacher training and research. Importantly, MSCF was now expected to broaden its attention to media for all physically disabled people, not only deaf people. At the 1969 symposium, Edward C. Carney, president of the Council of Organizations Serving the Deaf, gave the impression that MSCF's services might undergo considerable enlargement in the near future. Yet, technology still remained so new to educators and other professionals that when the law committee of the American Athletic Association of the Deaf met in 1969, *The Deaf American* magazine reported Carney's borrowing of an overhead projector from MSCF as an "innovation."[39]

## THE FIRST DECADE COMES TO AN END

In the years that he had served as the head of Captioned Films for the Deaf (later renamed MSCF), John Gough provided consistent leadership in not only bringing captioned films to deaf people in general but also in helping to shape national directions for the education of deaf students. He created a new perspective, which led to the retraining of teachers and the use of much higher quality educational materials that placed a strong emphasis on visual learning. By sponsoring the introduction of modern techniques that influenced generations of young deaf people, Gough effected one of the most significant changes in 150 years of deaf education.[40]

Gough retired on January 24, 1969. As the prime mover behind CFD and MSCF, he never lost sight of the adult deaf population and the original purpose of captioned films. Funds to caption entertainment films had increased nearly fivefold, resulting in outdated Hollywood features giving way to modern films released in color.[41] Now, as the 1960s came to a close, approximately fifty recently released films were being captioned each year and were made available to more than 1,000 groups of adults.

Gough looked back at his decade of leadership with great pride. He always recognized the contributions of the people who set the stage for his work.[42] He had seen the impact of automation and had made provisions for specially prepared captioned materials for training and rehabilitation centers. *The Deaf American* magazine printed the following about him: "[John A. Gough] was that rare man who managed to steer his singular vision through a maze of bureaucratic shoals during the turbulent sixties and bring it through intact—no mean feat when one considers the number of other federal programs that foundered during the same period or else had their identities swallowed in the corporate maw."[43]

From the start of the program, Gough knew that much more than captioned films were needed in the education of deaf students. When he first began working for the Office of Education, he had leaned heavily on his deaf friend Boyce R. Williams from the Office of Vocational Rehabilitation, who had been behind the original CFD legislation, navigating it through Congress. Gough had strongly advocated for the use of films to expand the experiences of deaf children and help them verbalize these experiences through printed captions. He saw how modern technology could bridge part of the communication gulf that presented a barrier to deaf people pursuing an adequate education.[44]

When Gough hired Gil Delgado to coordinate the Ball State workshops, he recognized that he had found a prize. Within a short time, in 1964, he expanded Delgado's role by making him his full-time assistant in the national office. Upon Gough's retirement in 1969, Delgado was first appointed acting chief of MSCF. Shortly after this he was promoted to chief. One of Delgado's priorities was the establishment of a fellowship program to train specialists in deafness and educational technology at the doctoral level.[45]

MSCF was progressing well. All of the schools for deaf children in the United States had some captioned film access in place, and ongoing teacher training made the use of the films more effective in instruction. Mac Norwood continued to perfect the systems and maintained contacts with universities and the schools.

## INSTRUCTIONAL TELEVISION TAKES HOLD

By 1969, MSCF had also produced 104 experimental half-hour television programs for deaf adults, thirty of which were available to educational television stations through the Electronic Resources Tracking System Library at Bloomington, Indiana. Schools and programs were gaining interest in television in the classroom, but the technology was still relatively new. That year, also, Edgar Lowell from the John Tracy Clinic in Los Angeles described some of the "tyrannies" of in-house productions of educational television, including the lack of compatibility of different types of video equipment, the lack of video technicians on the staff to care for the equipment, and teachers unprepared for the rigorous demands of producing a shooting script.

"Despite these efforts," Norb Nathanson told an audience that year, "we find ourselves—sixteen years after educational television began—just beginning the exploration and evaluation of television for the handicapped." He recommended that the Bureau for the Education of the Handicapped in the US Office of Education be the organization to provide the essential planning, direction, and fiscal support. He suggested that techniques developed for deaf people can have applications in other areas, and that deaf students could profit by utilizing techniques that had proven successful in other areas of special education. Nathanson presented an outline for such a plan, including teacher training, parent training, education of the general public, and student instruction.

The advancement of instructional technology, especially instructional television, was emphasized throughout the Sixth Symposium on Research and Utilization of Educational Media for Teaching the Deaf, held in Lincoln, Nebraska, March 16–18, 1970. Topics covered included media-based interaction, a survey of reactions of hearing individuals to captioned television, and programmed captioned movies as a supplemental medium for language development. The use of videotape was discussed in areas such as teacher preparation and self-evaluation, high school science and mathematics, and a captioned form of *Sesame Street.*

During this time in history there were many experimental ventures in the use of media occurring in residential schools. Deaf children were developing daily school news programs. With staff help, some schools had the children producing video segments for classes. Teacher training videos were developed, and there was a growing interest in teachers taping programs off air and having them available as additional media materials.

At the 1970 Nebraska symposium, MSCF Chief Delgado gave the welcoming presentation to more than 200 highly motivated participants who were eager to exchange ideas with well-informed specialists.[46] He mentioned that the symposium had been previously recognized in the field of educating deaf students, and it was the high point of the year for the Midwest Center and for MSCF. He also reminded the audience that a recent amendment provided for a resource called National Center on Educational Media and Materials for the Handicapped. There was one quote that Delgado felt was especially germane to the panel he was moderating on MSCF and instructional television: "The establishment in lower and secondary education is probably the most entrusted in the entire world. They are still teaching children as we were taught thirty years ago."[47] And in his own remarks, he emphasized, "A child today who comes into kindergarten has had from 3,000 to 4,000 hours sitting in front of that television tube, absorbing unstructured data that takes him way past Dick and Jane. And the system just doesn't respond to that."[48]

In "Television's Neglected Strengths," Edward Palmer, the director of research for the Children's Television Workshop, which had been producing *Sesame Street* since July 1968, emphasized the instructional relevance of film and television and that it should be heartening to those involved in teaching deaf students that a new look was beginning to characterize general-purpose instructional television.[49] The importance to the education of deaf students was in the increasing role that visual elements play in transmitting the meaning of the content. Palmer mentioned that it was unfortunate that this trend would always fall short of providing general purpose materials that are fully adequate to the needs of deaf students, but it may substantially increase the number of available materials suitable for captioning or for other forms of augmentation. He believed it would be fruitful to devise systematic procedures for instructing deaf students in visual literacy—that is, in "reading" the visual conventions of film and television. Teachers recognized the need to design instruction around the functioning senses of the child by employing a variety of visual techniques in their teaching: pictures, motion, cartoons, color, and so forth. Palmer made suggestions for when materials were in a documentary format to be supplemented with captions, including, importantly, supplementing captions with pictures, adding captions to enhance further interest to the picture, and having the captions help explain and even interpret what is being seen.

William Jackson, from the Southern Regional Media Center in Knoxville, presented statistics from a survey he had conducted. He found that by 1970, sixteen out of thirty-three schools reported having full-time media specialists, seven had a full-time technician, and four had both positions filled.[50] With MSCF's national funding efforts through the four media centers and the Ball State workshops, there had been an exciting increase in the number of schools hiring specialized personnel as well as obtaining equipment. Federal monies had provided primary support for television installations in twenty-three of the schools.

The first decade of the CFD program was one of great excitement for deaf people, who had been largely isolated from the culture of hearing people as it was portrayed in entertainment films. Deaf children were slowly being introduced to the culture and history of society through captioned educational films as well. However, the technology was not yet developed to level the playing field with hearing citizens. Access remained a tedious process. Theaters rarely showed films of any kind with captions. Television had yet to present any meaningful form of captioning. Social clubs for deaf adults, as well as school programs, continued to wait long periods of time to enjoy particular films.

# 9
# A Clamoring Demand

One development that did much to ensure a sense of success with captioned films and television was the appointment of Mac Norwood to assistant chief of Media Services and Captioned Films (MSCF) on May 3, 1970. With this promotion, Norwood became the highest ranking deaf professional in the Office of Education. The Deaf community knew that Mac possessed the skills and knowledge to help carry out in full the mandate from Congress to ensure equal educational opportunities for deaf people. He had been with CFD almost from the time it became federal law more than a decade earlier.

As Mac transitioned into his role, there was still much to do to continue the momentum. Selecting and captioning a film remained a tedious process. A summer month-long workshop in Montana with twelve teachers from eight states, for example, resulted in educational films and feature-length films being captioned by the participants.[1] But at this time in history, it still required about nine months after script revision to get a film in general circulation, and the amount of time varied with the length of film, grade level, and age group.[2]

Gil Delgado had become increasingly involved in the graduate school program at Gallaudet University, and he decided to work there directly to develop higher education and doctoral programs. When Powrie V. Doctor, a professor in the government department at Gallaudet, who was chair of Deaf Studies and also responsible for college scholarships, passed away while attending a World Federation of the Deaf Convention in France, Delgado was selected to fill Doctor's position and resigned as head of MSCF. On August 1, 1971, he began his new appointment as dean of the Gallaudet College Graduate School. This included the responsibilities of assisting departmental chairpersons and faculty to improve current graduate programs, especially in the area of counselor and leadership training.

In December 1971, Mac Norwood was just beginning his tenure as the program leader when he stood in front of an audience at the National Conference on Television for the Hearing Impaired, held in Knoxville, Tennessee, and presented a keynote address. He read the following letter, one of many received by the Office of Education encouraging the government to support captioning of commercial television:

> Dear Sir:
>
> We are a family with an eleven-year-old deaf child. There are few TV programs that he can enjoy as it is very difficult to read the actors' lips. Karl would benefit from captioned documentaries because these programs would help him with his school work. Also, he would be a part of the family when we are enjoying a movie or special show. Sponsors should find that this programming will reach new audiences; in addition, the FCC advised that you show captioned news.
>
> When will you start doing this?

Norwood paused. He was particularly moved by this letter. When he was Karl's age, he, too, had dreamed of having access to television shows. Now, at the time this conference began, he knew that the National Bureau of Standards was finalizing a project that was leading to the development of technology to provide captions for network television without "intrusion"—that is, imposing the captions on those who would rather not see them. He and many other advocates in the American Deaf community had been encouraging organizations to collaborate in the national captioning effort. Estimates of about 20 million deaf and hard of hearing persons in the United States comprised a significant minority group. The efforts in Little Rock, Arkansas; Jackson, Mississippi; and other locales to broadcast local newscasts interpreted with sign language on television were commendable, but many potential viewers such as the elderly did not know sign language and the news remained inaccessible to them. Experiments with captioned shows such as *Sesame Street* and the *Electric Company* were showing promise, and although there were many complaints a decade earlier about the reading speed of captions of 120 words per minute, a higher rate of 144 words per minute for all films aimed at adult audiences was now accepted.[3]

The technology for captioned television, however, remained unpredictable at the time of this conference, and Norwood wondered whether it might be possible for deaf people like young Karl to enjoy even captioned reruns on television, unless there could be, as he said presciently, "some method of modifying the receivers so that captions will appear only on sets owned by deaf persons."[4]

Indeed, access to commercial television was a dream the Deaf community had harbored from the very beginning of CFD. Mac Norwood was working with John Gough in CFD as early as January 1963, when the column "Film Fare" was published in *The Silent Worker* magazine. There, the potential of television was raised by a deaf writer—a hope "to see the day when there may be some Captioned Films."[5] "[Captioned commercial television]," the writer stated, "would help greatly to keep us abreast of the times. It would enable us to achieve a greater feeling of belonging to the world in which we live. This we wish very much to do."[6]

In 1971, those priorities were rapidly evolving—not only to provide entertainment access but also to meet the goals of enhancing academic success and improving assimilation of deaf people in American society. Norwood had confidence that a breakthrough in television access was slowly approaching. Over the past few years, he had seen the first telephone calls made between deaf people who used the mechanical teletypewriters along with a modem designed by a deaf physicist named Robert H. Weitbrecht in collaboration with Andrew Saks and James C. Marsters.[7] Fred Schreiber, the executive director of the NAD, was encouraging him to install one in the Office of Education. In addition, three years earlier in an article published in the *American Annals of the Deaf* titled "The Second Decade," Norwood predicted that the development of captioned commercial television would become a promising new offering. He ardently emphasized that captioned newscasts could be very important to deaf people, especially when an emergency notice of any type is given over the airwaves.[8] Throughout the 1960s, he and John Gough advocated for technological research to make this happen.

A year earlier at the 1970 Nebraska symposium, William Jackson from the Southern Regional Media Center had raised this issue with a very important request for the participants. Although the primary focus at the conference was on instructional television in the classrooms, Jackson encouraged his colleagues to actively advocate for captioned commercial television as it was, in his opinion, a very important need for both deaf children and adults. He saw an extension of the need, which in 1958 instigated the beginning of captioned films, and efforts by the deaf community to reach the commercial stations were already underway. "Let's start a clamoring demand for the rights of the deaf population that will bring about captioned public television," he said.[9]

With John Gough and Gil Delgado no longer serving in the national MSCF office, Norwood greatly depended on his staff as he began a new and critical period in this history of captioning. As he told an audience at the 1971 Nebraska symposium, the departure of Delgado from the program was a great loss. "Gil filled Dr. John Gough's shoes well by continuing to carry out the goals and objectives of the program. It was my privilege to have had the honor to work with both of these men whom we all hold in respect and affection."[10]

Norwood was a model for encouraging other deaf people to pursue leadership in this exciting period in history when both telephone and television technologies were advancing rapidly and activism among Deaf people was needed to empower the community. At the International Congress on the Education of the Deaf conference in Stockholm that year, he explained that the Council of Organizations Serving the Deaf in the United States was composed of sixteen organizations of and for deaf people, all of which were national in scope and which covered the spectrum of activity from research on vocational and social issues related to deaf people in athletics. "More and more the acceptance of responsible deaf leadership is resulting in positive services for the deaf community in the United States," he said. "I look forward to the day when this acceptance is acknowledged throughout the nations of the world so that the deaf community and the hearing community can function together on an equal basis to provide for a richer and better life for all of us."[11]

Norwood approached his new role in MSCF with much energy and enthusiasm. He had been a strong proponent of the use of visual media throughout the phenomenal revolution, and MSCF was having a continuous positive impact in the education of deaf students. Along with Gough and Delgado, he had spearheaded many enhancements that led deaf children to learn independently. Now, Norwood was prepared to build on the past success with a range of priorities—what he termed "new horizons."[12] He was confident that with the federal as well as grassroots support that had been an important part of the past decade, MSCF would continue to thrive and lead to a better education for a wider range of children with disabilities as the Department of Education was now advocating. This notion of addressing the educational needs of children with other disabilities was a special point Mac wished to make to his audience as he began his term in the national office. Although emphasis on the cultural and educational needs of deaf people would be continued, he would now need to oversee efforts to extend the instructional benefits of media to *all disabilities*. This would hopefully be addressed through the Special Education Instructional Materials Centers and Regional Media Centers Network. Through this system, stronger relationships between special education and instructional technology would be created so that educators of children with disabilities and technologists would have the opportunity to develop and validate new techniques. Basically, the approach would be to assess the needs of the children and follow this with the development of specific educational tools.[13]

Mac continued deep involvement in all phases of MSCF, supervising caption writers and the intricate business of adding captions to films, making frequent field trips to share and obtain ideas on needed film media and promoting their use by teachers, and overseeing the creation of lesson guides, manuals, and other material used to supplement the films. Among those he trained were Mike Waugh, a certified teacher who had moved from the Kansas School for the Deaf to Johnson County Community College (Kansas) as television director, to direct summer educational film and video captioning, and Bill Stark, Illinois School for the Deaf media director, to direct lesson guide production. Norwood also had organized the system of depositories and libraries used to distribute the films around the country.[14]

## BUMPS IN THE ROAD

At the conference in Knoxville in 1971, Robert Root from HRB-Singer, Inc. of State College, Pennsylvania, reported on Norwood's research into possible techniques for implementing captioned commercial television. According to Norwood, the expression of interest by deaf people and their enthusiasm with captioned films led many of them to explore the possibility of placing captions on TV screens and to examine whether the hearing population would accept this idea.[15] One experiment in Pennsylvania discovered a principal drawback to be the reaction of the general television audience to onscreen captions. In the study, 552 families were asked to view two captioned cable TV shows, *Bear Country* and *Big Red*. Thirty percent of the respondents said that captioning on *Bear Country* bothered them, and 24 percent said that captioning was bothersome on *Big Red*. Forty-seven percent of respondents were favorable to having open captions on selected TV programs but not on all programs. Forty-three percent were neutral and 10 percent were not in favor of onscreen captions. Root told his audience that these results suggested that an evolutionary approach would be needed to introduce captions on commercial television. He described how cable television systems were recently required to show programming that originated on local networks but that was passed through their systems. In order to convince television networks to caption their programs, additional data and statistics would be required.

The work of the government-sponsored MSCF with the motion picture industry during the 1960s also provided insight into matters that would have to be faced in exploring the new frontier of captioned television. For example, foreign movies were generally dubbed rather than captioned because many hearing viewers did not like subtitles. The fact that decisions in the media industry were often based on market preferences had to be considered.

Although Norwood was encouraged by the generally positive response to television captioning that had been revealed by the study, commercial networks were intensely competitive. The networks would likely not risk losing even a fraction of their hearing viewers. Again, firm data on the number of deaf and hard of hearing people who would be viewing captioned TV programs were needed. The networks, for example, would likely extrapolate that the 10 percent who reacted unfavorably to captioned television would be about twenty million viewers, which was larger than the deaf population. This was based on an estimated hearing television audience of 200 million at that time.[16]

It did not seem possible that the networks would compensate for any loss of hearing viewers. This conclusion intensified the determination to pursue forms of closed captioning technology that would permit captions to be displayed only on the screens of deaf and hard of hearing viewers who had decoders.

Open captions, those that appear on all television sets, can be viewed without any special equipment. Previously, occasional news bulletins, presidential addresses, or programming created by or for deaf and hard of hearing audiences were open captioned; that is, they could be viewed without any special equipment. Visionaries like Norwood were hoping to find a way for closed captions to be broadcast the same way as any other TV program is broadcast, but they knew that a device would be needed to make the captions visible on the screen. Experiments like the Pennsylvania study provided further incentive to explore "hidden" or closed captions.

At the same time, there was a growing concern in the Deaf community in reaction to Congress changing the mandate to MSCF to serve all people with disabilities. In particular, there were fears expressed that fewer entertainment titles would be available for deaf people, that there would be a greater percentage of educational films, and that inclusion of groups with other disabilities would result in a "watering down" of services to the adult deaf population. In addition, MSCF was providing far more than entertainment and educational movies. The program was also offering such educational media as filmstrips, overhead transparencies, and fingerspelling kits (loop cartridges), and grants had been made for scores of projects not involving visual media.[17]

This concern about entertainment films was temporarily abated when the Deaf community received promises that there would remain a steady flow of new titles through the MSCF distribution efforts.

## INTRODUCING NEW ADVANCES IN CAPTIONING

The exciting conference in Knoxville in December 1971 was a first major event in the evolution of captioned television that would unfold over two decades. It brought together many people who were interested in making television accessible to deaf and hard of hearing individuals. Participants also included parents, producers, representatives of federal agencies, and, significantly, representatives of the major television networks.

Several sessions of the conference were devoted to demonstrations of captioned television techniques. One interesting demonstration was given by the National Bureau of Standards and the ABC television network. A special device, which became known as a "decoder," was used to transmit printer captions across the television screen. This demonstration proved the feasibility of the closed caption system, which permitted captions to be seen only by a viewer who has special equipment to connect to the television set.[18]

Two possible technologies for closed captioning television were demonstrated. One system, developed by the HRB-Singer Company, transmitted the captions in a portion of the video signal that was normally received off the edge of the picture. One of the major networks criticized this method of presentation on the grounds that it might intrude on hearing viewers' pictures due to variations in reception, and that use of television sets to receive captions in this manner might reduce the life of the picture tube. Another system, originally developed by the National Bureau of Standards to carry time and frequency information, transmitted captions in a portion of the video system that was unused in normal transmissions. This portion of the signal is known as the "vertical blanking interval." The system was exhibited with the cooperation of ABC Studios in New York with a captioned presentation of the TV program *Mod Squad*.[19]

The development of the decoder was clearly a breakthrough. The National Bureau of Standards and the ABC television network had demonstrated that a portion of the television signal not used to show picture information could carry electronic codes (captions). A TV picture signal consists of 525 lines that flash their patterns on the screen thirty times a second. Among these lines are a few that are unused, and one of them was assigned to carry the closed captioned data. Decoders hooked up to the TV receivers interpreted the data and generated visible characters on the screen. The system developed using only the twenty-first line of the vertical blanking interval became known as the "line 21 system."[20] By December 1971, the National Bureau of Standards had verified that this technology could be used to provide captioning, and approximately one year later PBS began work to make the system a reality.

## PURSUING CLOSED CAPTIONING ON TELEVISION

Within a month, in January 1972, Norwood's staff met with a subcommittee of the National Association of Broadcasters to discuss the feasibility of captioned television. By June the subcommittee agreed that closed captioning was technically feasible, but it reported that certain steps had to be taken before such captioning could become a reality. First, the development of an effective decoder was needed, as well as extensive field tests of the system, including television receivers equipped with decoders.[21] Second, there should be an analysis of the cost-effectiveness of such a system, the length of time required to caption a TV program, the cost of captioning a program, the cost of equipment to broadcast, and the cost of the decoder to receive the signals.[22] The federal government contracted with PBS to do a feasibility study while ABC continued to support the project from its inception by providing technical assistance and advice as needed.

Then, the special decoding device featured at the Knoxville conference was attached to the closed circuit television system at Gallaudet College in Washington, DC, for another demonstration. The captions were provided by the Office of Education's MSCF, headed by Norwood.[23] Gallaudet president Edward C. Merrill began the demonstration with a brief welcome. He explained that the 5,000 graduates of Gallaudet "have a right to enjoy all programs and, especially, to get the news of emergencies—of earthquakes, tornadoes, etc."[24] Lewis M. Branscomb, director of the National Bureau of Standards, spoke about the previous captioning demonstration in Knoxville and explained that ABC provided the 16 mm film of the show *Mod Squad* and a script permitting captions to be made in Boulder, Colorado. The captioning was sponsored by the Office of Education. Present at this demonstration at Gallaudet were representatives from ABC, the National Society of Broadcasting, the FCC, WMAL TV, the NAD, the Council of Organizations Serving the Deaf, and the press. The secretaries of Health, Education, and Welfare (HEW) and Commerce and their staffs were also invited. The result was further realization that closed captioned television for deaf people was not only possible but also that it would likely become reality.

There was much enthusiasm about the captioning breakthrough among the Gallaudet students who viewed the broadcast. The Knoxville and Washington, DC, demonstrations of closed captioning provided numerous opportunities for informal discussions among deaf people, representatives of deaf organizations, parents, and television industry representatives.

MSCF then provided funding for the initial research into TV captioning and the development of the first decoders. Norwood arranged for HEW's Office of Education to contract with WGBH in Boston to produce an open captioned program and further develop the decoder. In 1972, WGBH created a team of engineers under John E. D. Ball. The Caption Center at WGBH, funded by HEW, was responsible for testing line 21. This testing included a fourteen-week experiment with Gallaudet researchers at twelve public stations.[25]

On an experimental basis, funding was provided to caption Julia Child's *The French Chef*, which was the first national broadcast of an open captioned program. It was aired across the United States on PBS on August 6, 1972.[26] Julia Child enthusiastically endorsed the captioning project, and her show was used to experiment with a variety of caption face styles and positions on the television screen.[27] The goal of this experiment was to gather additional data on both the reaction of the general viewing public to open captions and the size of the audience that captioned television could be expected to reach. In addition, there was a half-hour captioned show of President Nixon's inaugural address for nationwide rebroadcast over PBS stations that year. It was a slow start but filled with promise.

## CELEBRITIES BEGIN ADVOCACY

In addition to Julia Child's endorsement of captions, the actress Nanette Fabray was very enthusiastic about the access issue. At the CAID conference in Indianapolis in 1973, she explained, "When I had trouble in school, I thought it was because I was stupid, when the truth was, I couldn't hear what was going on."[28] She was completely deaf by the age of twenty, but four surgeries helped her to recover her hearing. She nevertheless remained an advocate for people with disabilities. As a member of

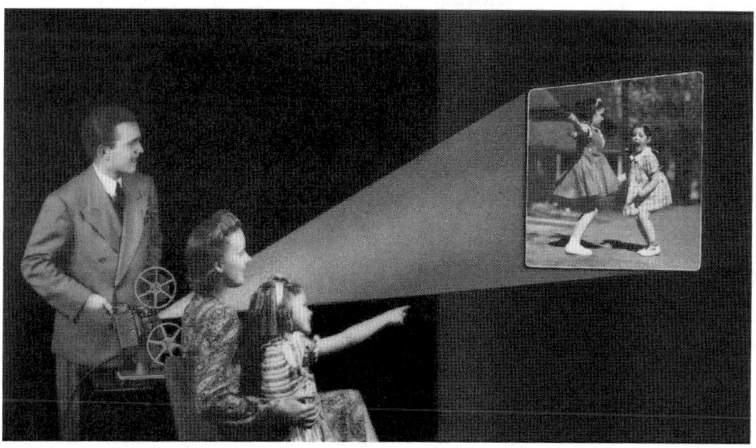

Watching a home movie.

the National Advisory Committee on the Education of the Deaf, Fabray stressed that there were many emergency situations that deaf people face, and captioning would be very helpful in saving lives in some situations. During the 1971 earthquake, for example, there were a thousand deaf families living below the dam in the San Fernando Valley. Radio, television, and police squad cars were broadcasting, urging evacuation, but most deaf people had no easy access to these messages. "Control of radio and television stations reverts to local management during emergencies," she explained.[29]

Norwood welcomed Fabray's advocacy of captioning, noting on June 18, 1973, that television had played a major role in serving the general population, "but its impact on the deaf has been minimal because there are few, if any programs, that can be understood."[30]

At 11:00 p.m. on December 3, 1973, *ABC Evening News* was rebroadcast on WGBH with open captions. It was subsequently made available to all PBS stations on August 6, 1974. It was seen late-night on more than 190 PBS stations (and aired for nine years, although it was not picked up by all stations). The *Captioned ABC News* was produced as the only timely newscast accessible to deaf and hard of hearing people at that time.*

* From Karen Peltz Strauss, *A New Civil Right* (Washington, DC: Gallaudet University Press, 2006), 208. The actual show on ABC was called *World News Tonight*, but with permission, WGBH shortened the title. Three other stations also carried the first broadcast. Within a few weeks, about ten additional PBS-affiliated stations around New England were included in the broadcasts. Jeff Hutchins, email to author, November 6, 2019. Hutchins, who later was a founder of VITAC, which became the largest closed captioning company in the United States, was the broadcast coordinator (supervisor) of a team of seven people hired in early 1973 to accept the challenge of captioning a twenty-two–minute program in just four hours. Jeff reported to Phil Collyer, director of The Caption Center at WGBH and producer of the program. When Collyer left, his replacement, Sharon Earley, appointed Jeff Hutchins as producer, a title he held until he left in 1980 when he joined the new National Captioning Institute (NCI) as director of systems development (essentially their chief technology officer).

# 10

## The Captioning Program Evolves

*To be shut off from television is more traumatic than words can describe.*

—Mac Norwood, quoted in Richard C. Levy, "TV Opens Up Line to the Deaf"

The rapid progress of the nation toward landmark legislation helped captioned television take flight. Throughout the 1970s, Mac Norwood's leadership guided the American Deaf community through the challenges of technological accessibility. But he was an educator at heart, and as one of the program chiefs in the Department of Education, he became immersed deeply in the mainstreaming movement as well.

In May 1974, during a major reorganization, the Media Services and Captioned Films branch of the Office of Education was elevated to division status within the Bureau of Education for the Handicapped and renamed the Division of Media Services. Two branches were formed within the division: the Captioned Films and Telecommunications branch and the Learning Resources branch. The major missions of the Captioned Films and Telecommunications branch were defined as (1) promoting the general cultural and educational advancement of deaf persons by acquiring, captioning, producing, and distributing captioned films and other media to deaf persons, and (2) promoting the educational advancement of deaf persons through the use of technology.[1] Both missions would involve research, production, and training activities in the area of instructional media. In addition to the *ABC Captioned Evening News* program, this branch funded the PBS closed captioning project, the National Theatre of the Deaf, and several other projects. The goals of the captioned films branch were clearly in harmony with the major government objective of making sure that educational services were provided to *all* persons with disabilities. The program activities would be directed toward ensuring that no citizen was denied the full range of motion picture films, television, telecommunications, and other technological advances that persons without disabilities customarily take for granted.[2]

**Use of Line 21 Delayed**

After PBS petitioned the FCC for permanent reservation for line 21 technology, there was a lengthy delay. In 1975, Karen Peltz Strauss was a supervising attorney at Gallaudet College's National Center for Law and the Deaf (later the name changed to the National Center for Law and Deafness).† She would be recognized as one of the country's foremost advocates on issues associated with telecommunications access by persons with disabilities. In her book, *A New Civil Right*, Strauss summarized that the major networks concerned during the testing of line 21 argued that closed captioning was "not yet technically possible."‡

Responding to the networks, consumers sought help from their congressional representatives. "It is an amazingly simple answer to a complex problem," the *Fort Collins Coloradoan* reported in its opinion column. "And yet, as of now, chances are slim to zero that it will soon become a reality. The Federal Communications Commission has the power to authorize caption use for Line 21, yet it is hesitating because of unanimous opposition from the four commercial networks."§ It was not until four years later, on December 8, 1976, that the FCC would allow broadcasters to *voluntarily* use line 21.

† Later, Strauss served as legal counsel for telecommunications policy for the NAD. Over a period of four decades, she coauthored and spearheaded efforts to achieve passage of several pieces of legislation requiring access to telecommunications, television, and the internet by people with disabilities. In the 1990s, she was appointed as deputy chief of a new consumer bureau at the FCC, where she spent a few years overseeing its disability and consumer access programs and policies, a role that she would later resume when invited to return to the FCC by the Obama administration in 2010. Strauss received her JD from the University of Pennsylvania Law School and her LLM from the Georgetown University Law Center, and an honorary doctorate from Gallaudet University. See more at http://tvworldwide.com/ittatc_bios/091101/karen_bio.cfm.

‡ Karen Peltz Strauss, *A New Civil Right: Telecommunications Equality for Deaf and Hard of Hearing Americans* (Washington, DC: Gallaudet University Press, 2006), 209.

§ "TV for the Deaf? Possible, but . . ." *Fort Collins Coloradoan*, October 5, 1976, 4.

## PUBLIC LAW 94-142

The Education for All Handicapped Children Act, or PL 94-142, is regarded as a landmark in the establishment of educational equity for school-aged children with disabilities in the United States. Its enactment in 1975 required that all children with disabilities three through twenty-one years old receive access to a public education in the "least restrictive environment." Children with disabilities were expected to receive special education and related services in elementary and secondary schools under the law's mandates.

PL 94-142 was unique in having the federal government provide funding assistance to the states to enforce the principle of educational equality for children with disabilities. It also established the conditions for state and local educational agencies to receive such assistance. States were required, for example, to adopt a policy to ensure that all children with disabilities could exercise their right to a "free appropriate public education" and to receive their education, whenever possible, with their peers who are without disabilities (often called "mainstreaming"). State and local educational agencies were expected to establish administrative procedures under which parents could question a school district's actions regarding the education of their children.[3]

The impact of this legislation was impressive, with more than four million children with disabilities served under PL 94-142 in the first ten years, and more than 400,000 education professionals trained to work with students with disabilities. Over 7.5 billion federal dollars were appropriated in that time.[4]

Mac Norwood was enthusiastic about the challenge of the new legislation. He said, "The time has come to pay greater attention to the application of technology to the problems of [students with disabilities] educationally, culturally, and socially. That is the promise of the new vision and the new vigor inherent in the PL 94-142 Amendments to the Education of the Handicapped Act." Building on the foundation laid in 1958 by the enactment of PL 85-905, Mac expected the future to see special education gain new strength through further technological advances.[5]

To Norwood and the many people involved with bringing access to the spoken word to deaf students through captioned films, the challenges presented by PL 94-142 seemed formidable. Chief among the concerns was how to reach deaf students enrolled in very small numbers in many public schools, how to train educators in mainstream schools, and how best to teach language-challenged students through the use of captioned media. When the law was first passed, Norwood worried about how communication would happen in mainstream classes. What could public schools offer deaf children in their native language? Would sign language be accepted? Would captions be welcomed in the classes? Norwood held discussions for years at the annual depository managers' meetings, from the year the law was passed in 1975.[6]

Throughout the years immediately following the passing of PL 94-142, Norwood stressed how the captioned films program had been the seedbed for a systematic, comprehensive service that was expected to benefit all children with disabilities. He also emphasized how telecommunications technologies were being developed as another viable means of bringing full services to citizens with disabilities. He believed that the changes in the educational system held great promise because we have "an unparalleled opportunity to bring about a combination of federal, state and local educators and industry working together for the benefit of all." An inclusive approach, he believed, would result in improvements across the board, as the advances in technology were applied to a nationwide effort. "Together we cannot lose."[7]

In support of PL 94-142, Norwood presented and published articles on numerous topics related to media such as "Audio Visual Communication and the Education of the Deaf in the United States"; "Media, Its Role in the Education of the Deaf in the United States"; and "Instructional Technology: Its Emerging Role in the Education of the Handicapped."[8] He discussed the federal perspective on the government as a change agent, and how the program offered services that directly and positively influenced the educational and social impact of media and technology on deaf people.

Related to this were new mandates from Congress that greatly expanded the scope of the program. Congress had authorized research, training, adaptation, production, acquisition, and distribution of educational media, not only for deaf people but also for those with other disabilities.

In an article published in *Exceptional Children* in 1976, Norwood explained that as a result of PL 94-142, his program became involved in helping to design educational media to help meet the learning and social needs of the overall population of children with disabilities, although those who were deaf or hard of hearing remained the chief target. As the two branches under the Division of Media Services worked together, they attempted to set up a systematic approach for enabling children with disabilities to acquire the skills they need through a comprehensive network of media centers providing a wide variety of services.

Norwood also reported that bookings of theatrical and documentary films to schools and other agencies and organizations serving deaf children and adults had reached 19,000 per month. Educational films for classroom use were distributed through a network of sixty film depositories and were seen by more than 1.5 million children annually. Consumer input in the selection and development of these films was considerable. Deaf and hearing classroom teachers, representing both public schools and residential schools, participated with media specialists and program supervisors in screening, evaluation, selection, caption writing, and the development of study guides.

## ACCESS TO ENTERTAINMENT FILMS

Despite the evolution of the captioning program and the explosion of media technologies and products, Mac Norwood was keenly aware, especially as a deaf person himself, that access to entertainment films remained difficult for most deaf citizens. Many towns and cities attempted to advertise the entertainment films to be shown locally with captions. In the fall of 1976, for example, the Solano (California) Community College Science Building Theater advertised the films *Start the Revolution without Me*, *Genghis Khan*, *In Cold Blood*, and others. The college encouraged all deaf persons as well as their families to attend.

Entertainment films selected for the captioned film collection were previewed by a cross-section of the deaf population through a contractual agreement with the NAD. Each reviewer provided an individual rating, and a consensus of these individual ratings would lead to recommendations to purchase the film. As with educational films, the consumer always played a major role in the selection process.

Throughout the 1970s, deaf people in local communities continued to gather in homes to watch captioned films together. David Strom recalled that his father, Jerry Strom, worked for the National Fraternal Society of the Deaf and subscribed to the captioning program. "We got a new movie almost every other week," David reflected. The Arlington Heights family, with both parents and three of the four children deaf or hard of hearing, always looked forward to the opportunity to view the open captioned films. Strom continued: "We were 'captioned movie central' for the local deaf/hard of hearing community and had anywhere from 5 to 20 people visit for viewing pleasure. Every once in a while, the bulb would burn out and we would wait impatiently for Dad to replace it before proceeding to finish the showing. At other times, the film itself would snap and break. Poor Dad would spend some time patching it up before we could enjoy the rest of the film."

After the film showing, David's younger brother was responsible for rewinding the film onto its original reel to be returned to the depository. He would then join the others in the kitchen to debrief about the movie over dessert and coffee. "DST (Deaf Standard Time) was alive and well then," David remembered, "when some of our guests couldn't get the hint they were overstaying their welcome after 2:00 a.m."[9]

Bernard Hurwitz, now a lawyer, remembered that in his youth, as a hard of hearing kid with no access to TV/movie popular culture, he nevertheless got to watch every important movie in the 1970s, thanks to his deaf parents and their friends who had a captioned film "exchange club" called the "Village Green Circle." He also recalled that he probably got to watch R-rated movies far earlier than his hearing peers because of it![10] Wanting to see the films so much, they bypassed restrictions on group size, as did several other deaf couples in Rochester. Each club had its own name and they always managed to return the films on time after exchanging them.

This practice was common. In the Chicago area, Jerry Nichols was part of a circle of deaf couples with deaf and hearing children who got together regularly to watch captioned films. "Due to our being restricted to getting films once a week, we conspired to exchange the films with each other before returning them to the Captioned Films center. That was how we got to watch 2–4 different films every week! That was something we will never forget."[11]

## NEWS SHOW FOR DEAF VIEWERS PRODUCED AT THE WGBH CAPTION CENTER

Jeff Hutchins, the broadcast coordinator in WGBH's Caption Center, worked closely with Mac Norwood to produce a news show for and *about* deaf or hard of hearing people. In removing commercials aired during the *ABC Evening News*, they gained six minutes that had to be filled. In order to make the show fulfill many of the news needs of this special audience, they dedicated two minutes to weather forecasts for the entire country, two minutes to reporting the results of professional and collegiate sporting events, and two minutes to features of particular interest to the Deaf community.[12] Those features were Hutchins's special area of interest, and he gave the captioning staff assignments for feature reports, including a regular feature called "Deaf Heritage." About once a year, the budget that Sharon Earley and Hutchins negotiated with Mac Norwood enabled them to do very special features. For example, they sent a crew to Morganton, North Carolina, to film deaf athletes who were preparing to go to the World Games for the Deaf. That way, when the games were held overseas, they had video and photographs of the American athletes that they could use to illustrate the results of the events. In 1979, Jeff was a member of a team of four people sent to Varna, Bulgaria, to observe the World Congress of the Deaf and send back timely reports each day with photographs of the day's highlights.[13]

The federal funding to PBS for the development of the system for closed captioning was separate from the funding allocated to WGBH (a PBS member station). The latter focused on open captioning of PBS programs and the *Captioned ABC Evening News*. Hutchins led that development, which required four distinct areas of invention. First, a method was needed to transmit data as part of a standard analog TV signal on any TV set that had been sold or would be sold in America (i.e., the work leading to the selection of line 21). Second, a system was needed to create the content of the captions and control their timing on the screen. Third, the content data had to be inserted into the signal of a prerecorded or live program. And fourth, consumer equipment would have to be able to detect, decode, and display the data found on line 21 without degrading the video display and make the captions readable.

Most of this development work was accomplished at PBS between 1974 and 1980.[14]

Issues to be addressed included the cost of captioning a TV program as well as who would supply the captioning. In addition, the commercial networks objected to the plan for WGBH in Boston to do the closed captioning, as they did not want to send their programs out to a rival network. The cost of captioning had also been a point of contention between the networks, and it was an issue during the FCC proceeding in 1976. The captioning cost was estimated by PBS at approximately $1,500 per hour, whereas network estimates were much higher. ABC then raised the idea of nonprofit, free-standing captioning, and this led to the proposal for a national captioning institute to be located in the Washington, DC, area where PBS-trained personnel and equipment would be useful.[15]

Norwood was especially busy in 1976. In addition to helping address these television captioning issues, he completed his doctoral dissertation at the University of Maryland. In his dissertation study, he compared an interpreted and a captioned newscast shown to deaf high school graduates and deaf college graduates. He found that deaf high school graduates and deaf college graduates receive significantly more verbal information from a captioned newscast than from an interpreted newscast. He reported that the educational level of deaf people appeared to determine the amount of verbal information received regardless of the mode of presentation used, but the greatest amount of information was transmitted by captions.

At the same time, progress was being made with recording closed captioned TV programs off air onto videotape with the hidden line 21 captioning, and the decoder developed by WGBH made the captions visible. Captions prepared before viewing time were made part of the tape, and the captioner edited the script, including editing out extraneous material. The new script was then typed and a "captioned" tape was created by playing the TV program on one machine and the captions on another, and then integrating the two into a new tape that contained audio, video, and captions.[16]

Despite the progress, there remained much to be done before captioned television would become a regular offering with a wider choice of shows. Norwood, always an educator at heart, saw the possibilities appearing brighter for deaf children and adults as each breakthrough occurred. As he continued to lead the national captioning efforts, he looked to the field for direction, priorities, and cooperation. "We still have far to go," he stated, "as we move toward the innovative approaches made possible by television, the utilization of computers, and the refinement of educational systems through design and development activities. This direction will undoubtedly lead us closer to the creation of a system which recognizes the problems of learning and their solutions."[17]

## NATIONAL CAPTIONING INSTITUTE ESTABLISHED

To overcome the objections of commercial networks to send their programs out to a rival network to be captioned, a new organization independent of WGBH was needed.[18] In the spring of 1979, Joseph Califano, secretary of HEW, announced the creation of the National Captioning Institute (NCI). John Ball, vice president of engineering at PBS, was NCI's first president (1979–1993). An electronics expert by training, Ball had considerable influence on the early captioning technology. When NCI was set up, he was the logical choice to become president and he served for fourteen years. His impact on the closed caption system was extraordinary.

The primary goal at that time was to figure out a way to create real-time captions during a live program and to invent a way to insert the real-time caption data into the broadcast signal.[19] Another important issue to be resolved at that time was the manufacture and sale of decoders and television sets with built-in decoding equipment. Following a lengthy exploration of manufacturing and distribution sales, which were generally negative, Sears, Roebuck and Company (Sears) was the only firm that consented to become involved. Texas Instruments had developed the circuits to translate the television signals into captions.[20] Sanyo, the company that manufactured television sets for Sears, would produce the decoders at their factory in Forest City, Arkansas, and Sears would market them through its catalog. In time, the decoders were made available from a variety of retailers.[21]

At first, the decoders didn't sell that well. In order to encourage the purchase of decoders, requests for proposals (RFPs) were issued to increase the number of captioned television shows available, to assist in reducing the cost of decoding equipment, and to encourage more private sector support. The result was an exciting blend of public and private support. RFPs were issued for closed captioned prime-time movies, specials, and regular TV program series; the manufacture of decoder chips (thus providing a subsidy to assist in reducing costs); and real-time captioning of national newscasts, closed captioned children's programs, closed captioned syndicated programs, closed captioned sports programs, and closed captioned local news programs along with seed funding for up to three years to encourage local community support for permanent captioning.[22]

> Jim House, too, was hooked on the new technology that brought him access to television. When he was involved in a medical experiment during the late 1980s, he had to go through pheresis, where a machine drew his blood, took out his white cells, and returned the red cells in his other arm. "It was a two-hour process I had to go through every month," he recalled. "Long enough for all of us in the room to watch a VHS movie. And, yes, I lugged a decoder box to the Red Cross blood center!"
>
> —Jim House, email to author, February 23, 2020

When NCI was formed, ABC and NBC television networks originally agreed to pay for (and broadcast) five captioned hours a week, whereas PBS would show fifteen. By September 1979, NCI was a functioning corporation in Falls Church, Virginia. Sears agreed to sell two types of decoding devices—an adapter, which attached to the TV set, and a nineteen-inch color TV with the decoding circuitry built in.[23]

John Ball and Jeff Hutchins lobbied hard for all the systems NCI developed to be in the public domain so that other entities such as The Caption Center at WGBH and the Canadian Captioning Development Agency could expand the amount of captioned programming available without having to develop their own systems.

Amy (Kielon) Stornello recalled how things changed for her in Rochester, New York, in the 1980s. Life was never the same for her. "After a decoder was installed in my parents' home, I could now participate in life! (Before, I'd be reading books all the time as TV would bore me.) Now with captioning, I had some fun!" During high school in the 1980s, her friends would go to people's houses to watch horror movies from Blockbuster. "I had to LUG the big brown box [decoder] to their house and figure out how to hook up the dang thing. It was a miracle when every NEW TV had built-in captioning decoders in them! HALLELUJAH!"

— Amy (Kielon) Stornello, email to author, April 14, 2019

"Too often the deaf are thrust out of the hearing world and have no choice except to socialize only with other deaf. But because of the captioning on the television, the barriers that separate the hearing and deaf worlds are broken down. Captioning gives us a common activity that hearing and deaf can do together."

—Deaf television viewer to NCI in Patricia Brennan, "National Captioning Institute," *Washington Post*, September 29, 1985

## THE SPECIAL MATERIALS PROJECT

There were other significant changes happening in the late 1970s. Len Novick, formerly media director at the St. Francis de Sales School for the Deaf in Brooklyn, was hired as the project's assistant director in March 1978. Then, in a time of change, in October 1979, Len succeeded Joe Domich as project director in a mutual agreement with Hugh Summers, executive director of CEASD.[24] That year, after thirteen years of managing distribution activities, the executive committee of CEASD joined with the executive committees of the CAID and the *American Annals of the Deaf* to approve the incorporation of a national office whose sole purpose was to maintain film distribution. The issues leading to this decision included the cost to maintain separate payrolls, and the need for a fair way to determine which of the organizations would bid for available contract or grant activities when there was an interest shared by more than one organization. In 1980, the name of the sponsoring organization managing film distribution was changed to the Association for Education of the Deaf (AED).

In response to a new Department of Education RFP, the AED renamed the Captioned Films for the Deaf Distribution Center, which had been established in 1966, as the Special Materials Project. The project, led by Howard M. Quigley, the first executive director of CEASD,[25] was responsible for several activities, including entertainment films, educational films, captioning, evaluation and selection of films, and scheduling and circulation through film depositories. The Special Materials Project also provided captioning workshops, lesson guide workshops, and materials for people with disabilities.

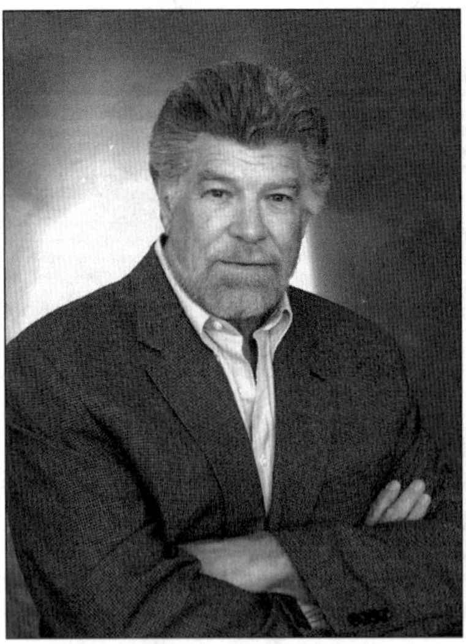
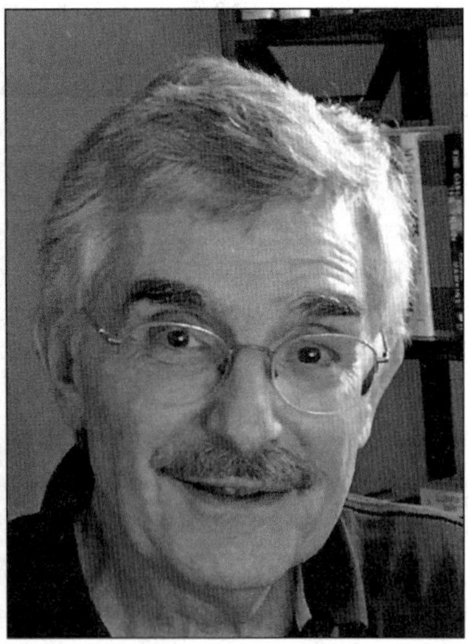

Under CEASD Executive Director Hugh Summers, Joe Domich, left, became the captioned films project director in 1976. Image courtesy of Joe Domich. In 1978, Len Novick, right, was hired as the project's assistant director. In October 1979, they switched roles. Image courtesy of Len Novick.

## FILMSHARE DEPOSITORY SYSTEM

At this time, CEASD initiated a study of the depository system for distributing films. This led to the recommendation to maintain all active depository sites, which served approximately 40 percent of the deaf and hard of hearing population and accounted for approximately half of educational captioned films in use at that time. Simultaneously, a change in the method of distributing materials among the depositories was also recommended.

In October 1979, the AED was awarded a two-year grant to develop an electronic interdepository loan system to assist in the management of the Educational Captioned Films program. The system was named FILMSHARE, and the plan was that each depository would help to fulfill the film requests made of any other depository in its region. The depositories were grouped into seven geographic regions. Each depository had previously served a registered school program in its geographic area, and schools with accounts were able to borrow film prints only from their designated regional depository. This necessitated having a copy of each film title in every depository. FILMSHARE had a great impact on the distribution system for educational captioned films. Before it began, the government had added approximately seventy new film titles to the Educational Captioned Films collection each year. The new policy allowed for the purchase of more than 150 titles per year, with fewer copies per title. Overall, more than 90 percent of all requests placed on the FILMSHARE bulletin board were booked.

A growing need existed, however, for ways to better manage the circulation of special education materials from distribution centers. The size of the film collections, and the addition of the inter-depository loan system, presented the Special Materials Project with the need to obtain accurate and timely data on film use and user demand. Novick recalled that this information was necessary for making recommendations to the federal government for print purchases and for the movement of prints from depository to depository and region to region.[26]

Staffing was a particular problem. The federal government, through its depository contracts, did not fully reimburse the operation of most educational captioned film depositories. The host institutions often subsidized depository staff costs and time. Many depositories did not have personnel available to confirm every booking. Bookings to users were automatically confirmed when a film was available. Computerizing this prevented increasing the workload of depository personnel. In addition, other automated procedures of the inventory control system cut down on the workload of depository personnel, enabling the operating costs to be kept low.

Throughout the 1970s, Salvatore J. Parlato was responsible for selecting and evaluating films. Parlato served as media director at NTID. In 1973, he authored *Films: Too Good for Words*. He also held a special interest in applying nonverbal films in the education of deaf children. Many of the evaluation and selection workshops were held at the Rochester School for the Deaf. Several multiday workshops were held each year, where teachers from schools for the deaf around the country would gather to screen and rate educational films for their suitability. At the end of the cycle, films recommended to the government that

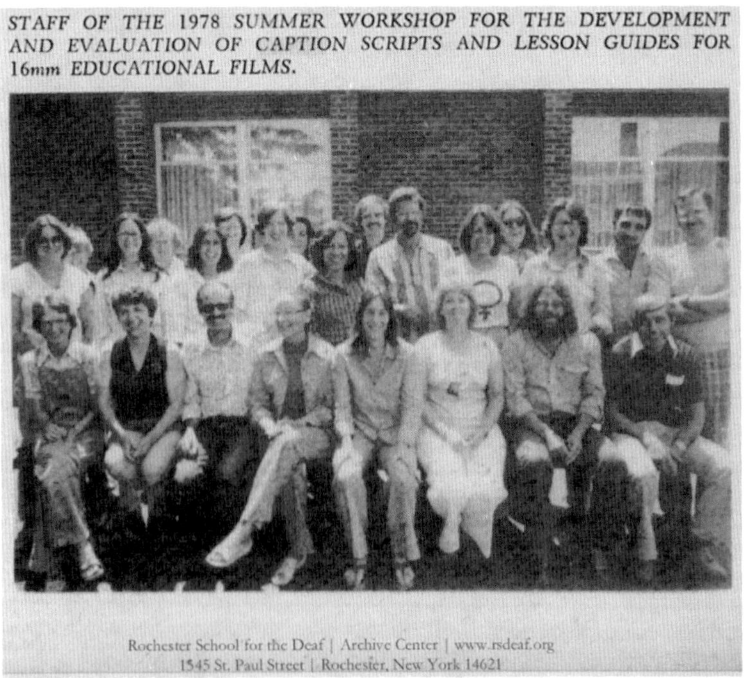


Staff and participants at the 1978 Summer Workshop for the Development and Evaluation of Captioned Scripts and Lesson Guides for 16 mm Films. Image courtesy of Gallaudet University Archives.

were chosen for purchase were captioned at a summer-long workshop. Lesson guides were also prepared, and these were distributed in bound volumes across the United States.

The Special Materials Project also initiated the Booking and Inventory Control System that operated on microcomputers, which were placed in each of the seven geographic regions to support FILMSHARE. The two microcomputer activities ultimately helped increase the use of educational films to deaf people around the country. When the system came online in 1981, it resulted in cost savings and overall efficiency.[27]

At the end of March 1981, the networks were broadcasting thirty hours a week. NCI reported that about 40,000 decoders had been sold that year. *Monday Night Football* and World Series games on ABC were also coded, as were all presidential speeches. Captioned advertisements were also part of the phenomenal growth during NCI's first year. At NCI, Barry Cronin explained that they did not plan on captioning commercials, but advertisers came when there were only 3,000 to 4,000 viewers. They were interested in both the commercial and the public service aspects. By 1981, there were more than 100 advertisers.[28] When asked whether there was satisfaction in being ignorant of the merits of this cereal or that toilet tissue, Cronin responded no. There was a thirst among deaf and hard of hearing viewers to know about everything that exists—good, bad, or indifferent. "They want the right to like and dislike ads and to decide what to buy."[29]

> According to DCMP's Captioning Timeline Highlights (https://dcmp.org/learn/25-captioning-timeline-highlights), the first live sports event with closed captioning was the Sugar Bowl on January 1, 1981. The game was played in the New Orleans Superdome, where the Georgia Bulldogs met the Fighting Irish of Notre Dame.

## MODERN TALKING PICTURE SERVICE TAKES OVER FILM DISTRIBUTION

In 1983, the federal government divided Norwood's program activity into two contracts: one for distribution and one for evaluation, selection, caption writing, and lesson guide production. As a consequence, AED kept its contract for film selection and captioning, but Modern

Talking Picture Service (MTPS), a for-profit company, became responsible for film distribution management. The MTPS headquarters was in New York, but the selection and captioning activities were primarily administered from offices in St. Petersburg, Florida. The movie and special interest depository was relocated to Florida, and the educational captioned film depositories remained in place in residential schools.

In 1984, the captioning program introduced videocassettes and was renamed Captioned Films/Videos for the Deaf (CFV). Then, in 1985, MTPS won the second of two contracts (for selection and captioning), having already won the distribution contract two years earlier.[30] A few years later, the two contracts reverted to being a single combined contract. MTPS was destined to manage the CFV activity

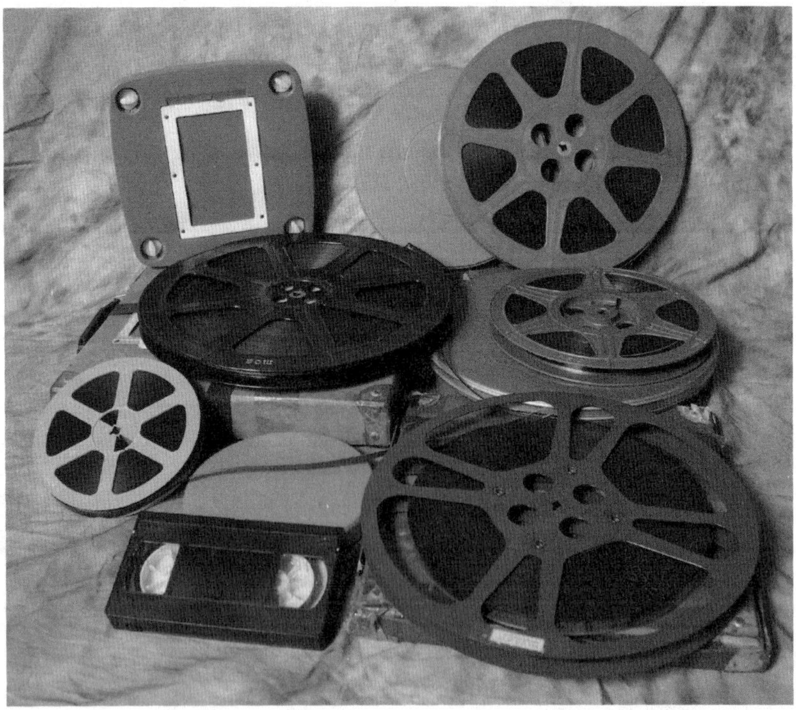

Sixteen millimeter films of various sizes ranging from 7 inches up to 15 inches in diameter. These heavy and bulky reels were replaced with smaller and lighter VHS cassettes measuring 4 inches by 7.25 inches (lower left), which helped save on shipping costs. Image courtesy of David H. Pierce.

for twelve years (1983–1991), overlapping with AED for two years (1983–1985). Don Zink and Mike Burke shared the management responsibilities for MTPS.

MTPS had no background in deafness or accessible media, and it turned the tasks over to subcontractors: Bill Stark was responsible for lesson guides for about a decade, and Mike Waugh for captioning. MTPS essentially trusted these leaders to run things in all respects from recruiting writers and training them to overseeing and approving final products. Bill and Mike refined the techniques and the training that Mac had begun years earlier. Stark, for example, added the first instructional graphics to the lesson guides, bringing graphic artists on board. He also experimented in different ways with presenting the "communication skills" component to the guides. Bill's wife, Dianne Stark, spearheaded a team of teachers who incorporated experimental instructional approaches into instructional guides for films. One of her guides for a social hygiene film was cited by an expert in the educational film industry as the best guide he had ever seen.[31]

CFV logo. Image courtesy of the Described and Captioned Media Program.

# 11

## The Battle for Captioned Television

To the millions of deaf and hard of hearing people in the United States, television before captioning had been "nothing more than a series of meaningless pictures."[1] In 1979, Tom Harrington, a twenty-eight-year-old hard of hearing audiovisual librarian from Hyattsville, Maryland, explained that deaf and hard of hearing people "would like to watch the same stuff as everyone is watching, no matter how good or how lousy. In other words, to be treated like everyone else."[2]

On March 16, 1980, closed captioning officially began on ABC, NBC, and PBS. The first closed captioned television series included *The ABC Sunday Night Movie*, *The Wonderful World of Disney*, and *Masterpiece Theater*. In addition, more than three decades after the movement to make movies accessible to deaf people began, ABC officially opened a new era by airing its first closed captioned TV movie, *Force 10 from Navarone*.[3]

By the end of March 1980, sixteen captioned hours of programming were going out over the airwaves each week, and by the end of May, Sears had sold 18,000 of the decoding units within four months of offering them for sale. Sears gave NCI an $8 royalty for each decoding device sold. The funds were used to defray the costs of captioning. In addition to building up a supply of captioned TV programs during its first year of operation, so that a sufficient volume would be available for broadcast, NCI concentrated on training caption editors. A second production center was established in Los Angeles and a third in New York City.

John Koskinen, chairman of NCI's board, reflected on the challenges the organization faced at this time. A much smaller market for the decoders was evident than that estimated through early surveys. As with the telephone modem that was simultaneously developing, the captioning decoders cost a significant sum for most deaf consumers in those days, and the expense of a decoder did not buy a lot because not all the captioned hours being broadcast were of interest to many people. Although the goal was to sell 100,000 decoders per year, NCI struggled to sell 10,000, and this presented a financial burden.

To help pay for the captioning costs, NCI also set up a "Caption Club" to raise money from organizations serving deaf people and from other private sources.[4] By December 1983, $15,000 was taken in and used to pay for subtitles on programs that otherwise would not be captioned.[5] By 1985, there were 3,500 members promoting the sales.[6]

Interestingly, when sales suddenly went up one year, NCI investigated and found that the Korean owner of an electronics store in Los Angeles was selling decoders as a way to enhance English learning.[7]

The next big breakthrough was the move toward the use of digital devices recently adopted by court recorders that, for NCI, allowed the captioning of live television. Having the ability to watch the evening news and sporting events with captions made the purchase of a decoder more attractive, as did the decline in its price over time.[8]

When the American television network NBC showed the twelve-hour series *Shogun* in 1980, thousands of deaf people were able to enjoy it. The $20 million series was closed captioned and 30,000 owners of the special decoder sets received the dialogue.

Jeffrey Krauss of the FCC admitted that deaf people had not had full access to television from the very beginning: "But by early 1980 it should be possible for the deaf and [hard of hearing] to enjoy many of the same programs we do via a new system called 'closed captioning.'" Sigmond Epstein, a deaf printer from Annandale, Virginia, felt that "there is more than a 100 percent increase in understanding." And Lynn Ballard, a twenty-five-year-old deaf student from Chatham, New Jersey, believed that closed captioning would "improve the English language skills and increase the vocabulary of deaf children."[9] Newspaper reports proliferated, describing the newfound joy among deaf people in gaining access to the common television. Educators recognized the technological advance as a huge leap forward. "I consider closed captioning the single most important breakthrough to give the deaf access to this vital medium," said Edward C. Merrill Jr., president of Gallaudet College, adding presciently, "Its usage will expand beyond

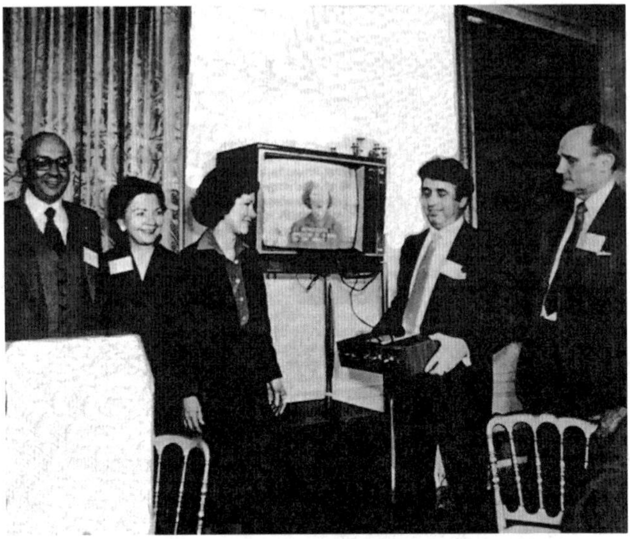

At the March 19, 1980, White House reception hosted by First Lady Rosalynn Carter, NCI President John Ball presented a captioning decoder ("adaptor") to the hostess. Looking on, left to right, NBC's Corydon Dunham, US Secretary of Education Shirley Hufstedler, and Sears's Robert Thompson. Image courtesy of Wallace Memorial Library at Rochester Institute of Technology.

the hearing-impaired."[10] And an ex-cop cried when his deaf wife wept for joy at understanding *Barney Miller*. He wrote a letter to the TV networks, cosigned by their six small children, to tell of the new world of entertainment and learning now open to his wife.[11]

> *3-2-1 Contact* was among the first group of television programs, and the first children's program, to be captioned in March 1980. This science education show produced by Children's Television Workshop aired on PBS member stations for eight years. Later that same year, *Sesame Street* became the second children's program to be captioned and became the longest running captioned children's program.
>
> —"NCI Recap'd," National Captioning Institute, http://www.ncicap.org/services/recapd/

The enthusiasm continued to spread swiftly among deaf people. Alan Hurwitz, then associate dean for Educational Support Services at NTID, and his family were all excited about the captioning of prime-time television programs. Hurwitz, who would eventually be president of Gallaudet University, was, like everyone else at this time, hooked on the new closed captioning technology. One of his favorite programs in 1981 was *Dynasty*, which was shown weekly on Wednesday night at 9 p.m. He flew to Washington, DC, early one Wednesday morning to meet with congressional staff members in different offices all day long. Not having a videotape recorder, he made sure he had scheduled a flight back home in time to watch *Dynasty*. After the meetings he arrived at the airport on time only to find out that the plane was overbooked and he was bumped off and scheduled for a flight the next morning. He panicked and argued with the airline clerk that he had to be home that night, and stressed that he couldn't miss the flight. He was put on a waiting list and there were several folks ahead of him. Then, when he learned that he would definitely miss the flight, he went back to the clerk and insisted that he get on the plane. He explained that he had no way to contact his wife and was concerned about his family. Finally, the clerk went inside the plane and asked if anyone would like to get off and get a reward for an additional flight at no cost. One passenger volunteered to get off and Hurwitz was allowed to take his seat. The plane left a bit late and arrived in Rochester barely in time for him to run to his car in the parking lot and drive home to watch *Dynasty*![12]

And even with the positive response from many consumers, it was reported in 1981 that the Sears TeleCaption decoders were not selling well. It was a catch-22 situation. "People hesitate to buy because more programs aren't captioned; more programs aren't captioned because not that large an audience has adapters."[13] Increasing one would clearly increase the other. The question was whether to wait for "the other" to happen. To do so would most likely endanger a considerable federal investment as well as the continued existence of the system. Some theorized that the major factors for the poor sale of decoders were the depressed state of the economy, the lack of a captioned prime-time national news program (which deaf and hard of hearing people cited as a top priority), insufficient numbers of closed captioned programs, and an unrealistic expectation by some purchasers that decoder prices would decrease in spite of the fact that the retailer markup was slightly above the actual production cost.[14]

## CAPTIONING A TV PROGRAM: A CONTINUING CHALLENGE

On average, it took twenty-five to forty hours to caption a one-hour program. First, the script was typed verbatim, including every utterance such as "uh," stuttering, and so forth. Asterisks were inserted to indicated changes in speakers. Next, the time and place of the wording was checked in the program. The transcript was examined for accuracy, noting when the audio starts and stops, and then it was necessary to decide whether the captions should be placed on the left, right, or center of the screen.[15] In 1981, NCI's goal was to provide no more than 120 to 140 reading words per minute for adult programs and sixty to ninety for children's programs.[16]

"We have to give time for looking at the picture," Linda Carson, manager of standards and training at NCI, explained. "A lot of TV audio goes up to 250 or 300 words per minute. That's tough for caption writers. If the time lapse for a 15-word sentence is 4 ½ seconds, then the captioner checks the rate computation chart and finds out she's got to do it in nine words."[17]

Carl Jensema, NCI's director of research, who lost his hearing at the age of nine, explained that at the start of kindergarten, hearing children have about 5,000 words in their speaking vocabulary, whereas many deaf children are lucky to have fifty. Consequently, deaf children had very little vocabulary for the school to build on. Jensema believed that closed captioning might be the biggest breakthrough for deaf people since the hearing aid. He was certain that a high degree of exposure to spoken language through captioned television was the key to enhanced language skills in deaf people.[18]

## CBS RESISTS

Although ABC, PBS, and NBC were involved in collaborating with NCI to bring captions to deaf audiences, the system CBS supported, teletext, was developed in the United Kingdom and was at least three years away from implementation. "It seems to me that CBS, by not going along with the other networks, might be working in derogation of helping the deaf or the hearing-impaired to get this service at an earlier date—and I don't like it." FCC commissioner Joseph Fogarty told Gene Mater, assistant to the president of the CBS Broadcast Group.[19]

Despite the success of line 21 captioning, CBS's Mater believed the teletext system was "so much better" and the existing system was "antiquated." "I think what's unfortunate is that the leadership of the hearing-impaired community has not seen fit to support teletext. Those people who have seen teletext recognize it as a communications revolution for the deaf."[20] In contrast, NCI's Jeff Hutchins summarized that the World System Teletext presented various disadvantages. It could not provide real-time captioning, "at least not in the way we have seen it . . ." Also, it could not work with home videotape. He believed that even if World System Teletext were adopted by the networks and other program suppliers, the technology would not be an answer for the needs of the American Deaf community. He also explained that "too many services now enjoyed by decoder owners would be lost."[21]

CBS even petitioned the FCC in July 1980 for a national teletext broadcasting standard.[22] Following this, the Los Angeles CBS affiliate announced plans to test teletext in April 1981. "CBS was so opposed to line 21 that even when advertisers captioned their commercials at no charge to CBS," Karen Peltz Strauss wrote, "the network allegedly promised to strip the captions off before airing the ads."[23]

CBS continued its refusal to join the closed captioning program, largely because of its own research into the teletext system and because the comparatively low number of adapters purchased. The NAD accused CBS of failing to cooperate with deaf television viewers by refusing to caption its TV programs.[24]

The NAD planned nationwide protests shortly after this. Hundreds of captioning activists gathered at studios around the country. In Cedar Rapids, one young child carried a sign that read, "Please caption for my Mom and Dad." Gertie Galloway was one of the disappointed deaf consumers. "CBS has not cooperated with the deaf community," she stated. "We feel we have a right to access to TV programs."[25] She was one of an estimated 300 to 400 people carrying signs, who marched in front of the CBS studio in Washington and who were asking supporters to refuse to watch CBS for the day. Similar demonstrations were held in New York, where there were 500 people picketing, and the association said that protests had been scheduled in the more than 200 communities where CBS had affiliates.

Harold Kinkade, the Iowa Association of the Deaf vice president, said, "I don't think deaf people are going to give up on this one. We always fight for our rights to be equal with the people with hearing."[26]

The drama increased in August 1982 when it was announced that NBC was dropping captions due to decreased demand. It was two years after NBC had become a charter subscriber. John Ball, president of NCI, said, "There is no question that this hurts. This was a major revenue source for NCI. I think the next six months or so are going to be crucial for us."[27]

Captioning advocates included representatives from NTID, the National Fraternal Society of the Deaf, Gallaudet, and NAD. Karen Peltz Strauss tells the story of Phil Bravin, chair of a newly established NAD TV Access Committee, who represented the Deaf community in a meeting with NBC executives. Although the NBC meeting was successful, CBS was still resisting and Bravin persisted. As Strauss summarized, "After one particularly frustrating three-hour meeting with the CBS President of Affiliate Relations Tony Malara, Bravin left, promising to 'see you on the streets of America.'"[28]

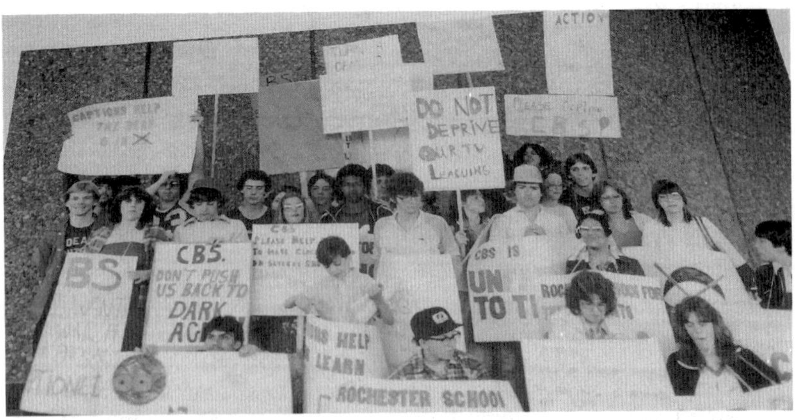


In 1982, Rochester School for the Deaf students joined the CBS protest. Image courtesy of the Paul and Sally Taylor collection at Southwest Museum of Engineering, Communications and Computation (Ed Sharpe, director). www.smecc.org.

In 1984, CBS finally gave in, and the network dual encoded its television programs with both teletext and line 21 captions. The issue with NBC also resolved, and by 1987 the network was paying a third of the cost of the prime-time closed captioning. The rest was covered by such sources as independent producers and NCI, with funds from the US Department of Education used for captioning on CBS and ABC as well.[29]

In his book *Closed Captioning: Subtitling, Stenography, and the Digital Convergence of Text with Television*, Gregory J. Downey summarized that because the film industry was unwilling to perform same-language subtitling for its domestic audience, the focus of deaf and hard of hearing persons' "educational and activist efforts toward media justice through subtitling in the 1970s and 1980s had decisively moved away from the high culture of film and instead toward the mass market of television."[30]

Meanwhile, teachers and media specialists in schools for deaf children across the United States were reporting that their students voluntarily watched captioned TV shows recorded on videocassettes over and over again. These youngsters were engaged in reading, with its many dimensions and functions. In the opinion of some educators, television was indeed helping children learn to read.

People at NCI looked forward to spin-offs from their efforts. They liked to point out that experiments on behalf of deaf people produced the telephone and that the search for a military code to be read in the dark led to braille. Closed captioning should be no different in that regard. The technology also showed promise for instructing hearing children in language skills. Fairfax County public schools in Virginia, authorized a pilot project to study the effectiveness of captioned television as a source of reading material. The study explored the use of closed captioned television in elementary classrooms, evaluated teacher and student acceptance of captioning as an aid to teaching reading, and served as a guide to possible future expansion of activities in this area. Instead of considering television as part of the problem in children's declining reading and comprehension skills, Fairfax County wanted to make it part of the solution. Promising results were found in this study as well as in other NCI-funded studies with hearing children, and when NCI's John Ball submitted his budget request to Congress for fiscal year 1987 he was citing "at least 1,500,000 learning disabled children" as a potential audience for captioning and the market for decoder purchases.[31]

In a personal tribute to Carl Jensema, Jeff Hutchins wrote that the only aspect of NCI that really made it an "institute" was the work Carl did to research many different aspects of captioning, including its readability and efficacy among consumers. His work led to a revision of techniques, which made captioning more effective. Once Carl left NCI and the research department was shut down, NCI was not really an "institute" any longer. John Ball also believed in the importance of Jensema's research at NCI. His studies clearly demonstrated the impact of captioning on NCI's important audience.[32]

## REAL-TIME CAPTIONING

As early as 1978, the captioning program began to fund developmental work in real-time captioning with the objective of making it possible to caption live programs, such as news, sports, the Academy Awards, and space shuttle launches. This developmental work, however, did not result in the system finally being used. The Central Intelligence Agency (CIA) was exploring a system that would allow the spoken word to appear in printed text. As it turned out, a private concern resulted from the CIA project, Stenocomp, which marketed computer translations to court reporters. The Stenocomp system relied on a mainframe computer and was thus too cumbersome. However, when Stenocomp went out of business, a new firm developed—Translation Systems, Inc. (TSI) in Rockville, Maryland. Advances in computer technology made it possible to install the Stenocomp software into a minicomputer. This made it possible for the NCI to begin real-time captioning using a modified stenotype machine linked to a computer via a cable.[33]

On December 20, 1982, the Ninety-Seventh Congress passed a joint resolution authorizing President Ronald Reagan to proclaim December as "National Close-Captioned Television Month." The proclamation was in recognition of the NCI service that made television programs meaningful and understandable for deaf and hard of hearing people in the United States.

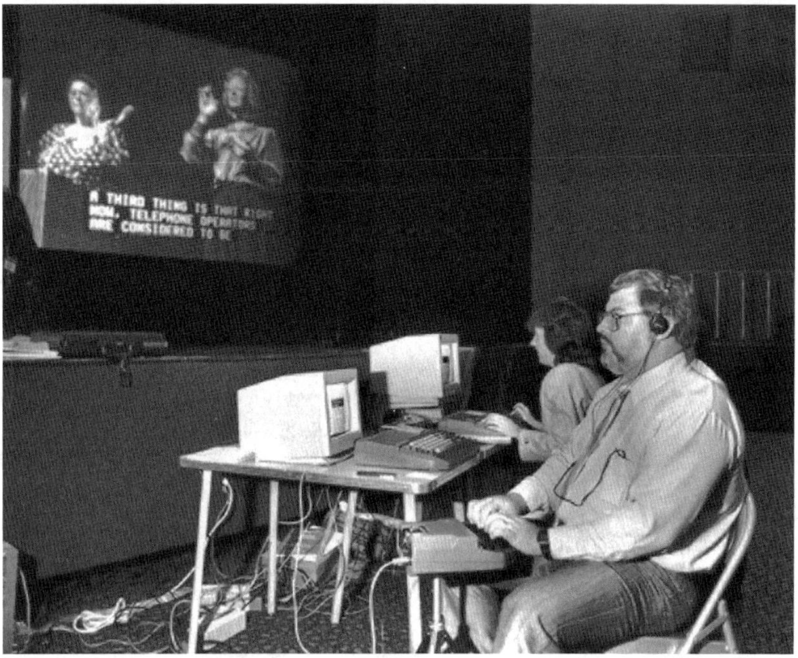


Real-time captioning is often provided at "live" events, such as conferences, ceremonies, speeches, and live broadcasts. Most often, the verbatim text is displayed on a computer screen, TV monitor, or a large screen. Image courtesy of Gallaudet University Archives.

By 1982, NCI was applying real-time captioning to a variety of televised events, including newscasts, sports events, and other live broadcasts, bringing deaf households into national conversations. The information, with correct punctuation, was brought to viewers through the work of stenographers trained as captioners typing at speeds of up to 250 words per minute.[34] Real-time captioning was used in the Supreme Court to allow a deaf attorney, Michael Chatoff, to understand the justices and other attorneys.[35]

However, fidelity was not the case for many years on television, and problems existed with real-time captioning. In real-time captioning, an individual typed the message into an electric stenotype machine, similar to those used in courtrooms, and the message included some shorthand. A computer translated the words into captions, which were then projected on the screen.[36] Because "this captioning occurred 'live' and relies on a vocabulary stored in the software of the computer, misspellings and errors* could and did occur during transcriptions."[37]

Over the years, many have worked toward error reduction in real-time captioning. As the Hearing Loss Association of America has summarized, "Although real-time captioning strives to reach 98 percent accuracy, the audience will see errors. The caption writer may mishear a word, hear an unfamiliar word, or have an error in the software dictionary. In addition, transmission problems can create technical errors that are not under the control of the caption writer."[38]

At times, captioners work in teams, similar to some sign language interpreters, and provide quick corrections. This was the approach the pioneer Martin Block used during the Academy Awards in April 1982. Block typed the captions while a team of assistants provided him with correct spellings of the award nominees.[39]

There has also been a growing body of educational research supporting the benefits of captions. As one example, E. Ross Stuckless referred to the concept of real-time caption technology in the early 1980s as the "computerized near-instant conversion of spoken English into readable print."[40] He also described the possibility of using real-time captioning in the classroom. Michael S. Stinson, another former colleague of mine and also a deaf research faculty member at NTID at RIT, was involved with Stuckless in the first implementation and evaluation of real-time captioning as an access service in the classroom. Stinson subsequently obtained numerous grants to develop C-Print access through real-time captioning at NTID, where hundreds of deaf and hard of hearing students have benefited in this postsecondary program. C-Print also was found successful in K–12 programs.[41]

Communication Access Real-Time Translation (CART) is another service provided in a variety of educational environments, including small groups, conventions, and remote transmissions to thousands of participants viewing through streaming text. Displays include computers, projection screens, monitors, or mobile devices, or the text may be included on the same screen as a PowerPoint presentation.[42]

---

* An example of a more serious caption error was observed during Ari Melber's interview on MSNBC with Dr. Anthony Fauci in July 2021. The caption read: "One statistic, Ari, that's really critical is that if you look at the deaths due to COVID-19 in the United States, 99.9% of them are among vaccinated people." This error was corrected on the internet later.

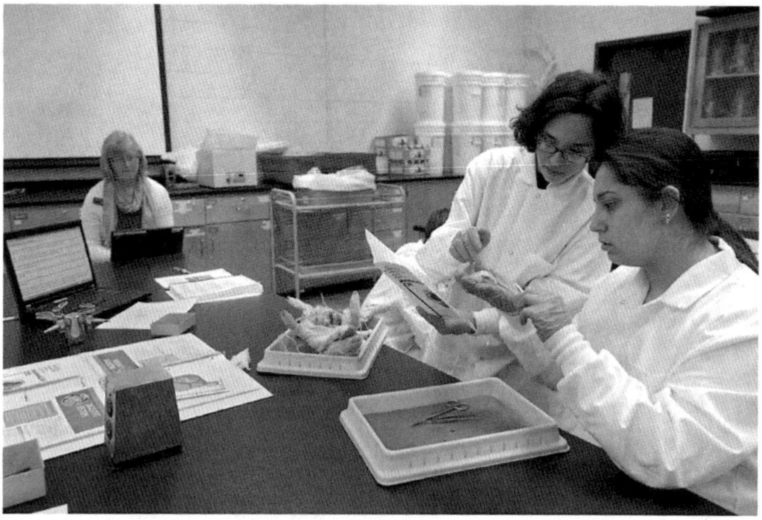

A C-Print captionist (background) types the instructor's spoken message in a biology class at RIT. The message is displayed on a laptop near the deaf student. Image courtesy of Rochester Institute of Technology. All rights reserved.

Special approaches have been used in educational environments. For example, at NTID, where C-Print was developed by Stinson, the scripts of the classroom presentations and communication between professors and students are printed out, and errors are corrected and given to the students to study.

In October 1984, ABC's *World News This Morning* became the first daytime television program to be broadcast to viewers with decoders through real-time captioning technology. Within a few weeks, the ABC's *Good Morning America* was broadcast with captions as well. "This is a major milestone in the evolution of the closed-captioned television service," John E. D. Ball declared, describing it as a "valued medium" to deaf and hard of hearing viewers. Don Thieme, a spokesman for NCI, explained that the Department of Education had provided The Caption Center with a $5.3 million contract. These two programs joined ABC's evening news program *World News Tonight* and the magazine show *20/20* as the only regularly scheduled news and public affairs available for deaf viewers.[43] The captioned news programs would be phased in gradually during the summer and early fall. Real-time captioning was also provided for the presidential political debates around this time. More than sixty-five home video movies had also been captioned for deaf people.[44] This was an important step toward providing more access to entertainment movies for deaf consumers.

The first time the Super Bowl was aired with closed captions was on January 20, 1985. In September 1985, ABC's *Monday Night Football* became the first sports series to include real-time captioning of commentary.[45] ABC, its affiliates, the US Department of Education, advertisers, corporations, program producers, and NCI's Caption Club helped to fund this program. Using stenotype machines, speed typists in Falls Church, Virginia, listened to the telecast and produced the captions at about 250 words per minute and they appeared on the screen in about four seconds. Each word was not typed separately. Instead, the captioner stroked the words out phonetically in a type of shorthand. Then a computer translated the strokes back into the printed word. These words were sent through phone lines to the ABC control room in New York City, where they were added to the network signal and transmitted across the country. Darlene Leasure, who was responsible for football, described one of the challenges she encountered: "When I was programming my computer at the beginning of the season, I found thirteen Darrels with seven different spellings in the NFL. It's tough keeping all those Darrels straight."[46]

As TV shows with closed captions grew in popularity, deaf people were attracted away from the captioned film showings at social clubs or other such gatherings.[47] The groups continued to hold their meetings, but for most gatherings the showing of captioned films gradually stopped.

At the same time, telecommunications advances had brought telephone access to deaf people and there was less need for face-to-face "live" communication. Together, the visual telecommunications and captioned television technologies profoundly impacted the way deaf people interacted.

## MAC NORWOOD RETIRES

Mac Norwood retired in 1987. For many years, almost since CFD's inception, he had worked in and/or directed the caption efforts. Along with answering countless letters from members of Congress, responding to concerns from the Deaf community, and serving as liaison officer for government projects such as NTID, Norwood pushed forward with more and better captioning of films, television, and videotapes. McCay Vernon, editor of the *American Annals of the Deaf*, wrote that deaf people all over the United States owed Norwood tremendous appreciation for what he had accomplished. "Every evening when they see television news, entertainment and educational programs with captions, it should be remembered that the one individual most responsible for that remarkable opportunity is Dr. Malcolm [Norwood]."[48]

Under Mac's leadership, the techniques that were perfected in the captioning of films during John Gough's term as chief began to find a place in network television, opening up to deaf citizens a previously restricted world of daily information and entertainment.[49] Like Gough, Norwood had done much to apply technology to better the lives of deaf people educationally, culturally, and socially. He had excelled in continuing Gough's legacy and, in this second decade, Norwood was the principal influence in helping deaf citizens attain "media equity."

For nineteen years, Norwood enthusiastically advocated for enhancing the education of deaf students through instructional media. At a presentation given at Gallaudet University in 1988, he stressed that the "name of the game today is educational services" and the need for continued research on the use of technology. He wrote of the upcoming development of learning resource centers, and he emphasized, especially, that "the utilization of television as a means of bringing deaf persons further into the mainstream of the general population also has a top priority."[50] Thanks to Mac's initiatives, closed captioning was making inroads into the educational arena.

Norwood also saw closed captioning as a service begin to catch the attention of the corporate and advertising communities. In 1988, there were more than 160 hours per week of captioned programs available on the network and in syndication. A natural adjunct of the closed captioning explosion was the home video arena, although it was still a crawl for home video distributors and stations to caption their tapes. Even though 72 percent of the decoder owners also owned VCRs, only about 1,000 home video titles from sixteen different companies were available. The National Video Clearinghouse listed more than 53,000 titles available on video from more than 1,000 companies.[51] Marlee Matlin explained that like millions of other Americans, she loved to watch videos at home, but many of those at her local video store were not closed captioned—not only then-current films like *Dirty Dancing*, but classic films like *The African Queen*, *All About Eve*, and the films of Capra, Hitchcock, and even her favorite, *The Wizard of Oz*. "Sad, but true," she signed. "There is a great deal of work to be done."[52]

Matlin remained optimistic, however. "I am extremely grateful for the progress that has been made. . . . You have made the lives of hundreds of thousands of Americans whole," she told the Corporate Advisory Council of the National Captioning Institute and representatives of companies such as Aaron Spelling Productions, Coca-Cola, the Kellogg Co., Warner Home Video, Walt Disney Studios, Home Box Office, and Showtime.[53]

## LEN NOVICK JOINS NCI

In 1986, Len Novick left CFV and joined NCI, where he remained until 1994. During his tenure there, he saw captioning services greatly expanded. Novick recalled, "We were able to successfully 'mainstream the technology' of captioning," he wrote, "which was developed originally for deaf persons, to an accepted feature available to all TV viewers in the United States."[54]

In 1987, the Department of Education began issuing RFPs for grant activities designed to increase captioned programming where there was little available to date, especially in the areas of national and local news, sports, and daytime programs. NCI was especially interested in cooperative funding arrangements with program providers, and Novick developed a proposal to the Department of Education resulting in funding owed to NCI from prior years of captioning.[55]

The benefits of captioning had been well established, including for hearing students. In one 1987 study sponsored by CBS and Boston University, Milton Goldman, a Los Angeles high school teacher, had found that "the reading comprehension and vocabulary skills of his hearing students improved considerably when they were forced to read captions rather than hear a program's audio track."[56]

NCI also commissioned studies revealing similar improvements. One of these studies, by Patricia S. Koskinen, Robert M. Wilson, Linda B. Gambrell, and Carl Jensema, titled "Using the Technology of Closed-Captioned Television to Teach Reading to Handicapped

Students" and published by the US Department of Education in 1987, showed that the addition of captions to conventional television presented benefits in sight vocabulary and comprehension skills of both students with learning disabilities and students who are deaf.[57]

"You're saving my sanity and probably my life . . . I was a teacher of languages, English, French and Latin, and a lover of all forms of communications. I've been totally deaf for a year . . . I get through the day because I know that for a little while at night I can watch TV and read your captions and forget there is a glass wall between me and the world."

—Deaf television viewer to NCI from Patricia Brennan, "Decoding TV for Hearing Impaired Opens Up a New Video World," *The Morning Call* (Allentown, Pennsylvania), October 1, 1985, 48.

In the 1980s, deaf people were making the best of using the heavy, bulky decoders. Captioned TV did not come Nancy Abbott's way until she was an adult and her family had to buy a "box."

Abbott recalls, "Deaf people who were involved in community activities would meet at night without worrying about making it home to view a program—simply because they couldn't follow the programs. When captioned TV came along, they were also the first to buy VHS recorders to enable them to watch TV with captions at a convenient time. *Dynasty* was a popular program, however, and we had to end meetings so we could get home to view that program.

—Nancy Abbott, email to author, December 22, 2018

## THE "DECODER CHIP BILL"

In 1988, the Commission on Education of the Deaf (COED), which was created by the Education of the Deaf Act of 1986 and chaired by Frank Bowe, made an important recommendation for legislation to mandate that all new television sets have circuitry for decoding and displaying closed captions. The COED report also recommended that federal legislation be enacted to require television programmers to caption their shows.[58]

The next year, NCI chose International Telephone and Telegraph to develop a single caption chip that could be efficiently built into every television. This effort was aimed at expanding the audience for captioned television and promoting universal acceptance of the service in the eyes of programmers. NCI simultaneously began working with Congress to have language enacted requiring TV manufacturers to have the caption decoder technology built into new TVs.[59]

The built-in decoder would hopefully eliminate the need for the separately purchased decoders for home television sets. At the same time, progress was being made with recording closed captioned TV programs off air onto videotape with the hidden line 21 captioning, and the decoder made the captions visible.

Karen Peltz Strauss writes of the resistance encountered when the FCC questioned the value of the decoder circuitry requirement: "This was hardly the first time that deaf consumers would not have the support of the FCC in their quest to expand telecommunications access. Fortunately, this time, the FCC's absence as a partner was offset by backing from a number of other important sources. Television programmers, eager for larger audiences, believed that a decoder bill would force television manufacturers to participate in the captioning effort."[60]

In March 1989, Frank Bowe was very concerned about the effects of decoder sales on the future of television access through captions. In an opinion piece published in many newspapers around the country, the deaf leader of the COED wrote, "My meetings with network executives lead me to believe that unless the chip, or something very much like it, achieves market penetration by 1992, captioning on network television will drop precipitously."[61]

Tom Harkin was the leading proponent of the new decoder legislation in the Senate in 1989. Harkin approached Bowe about meeting with the foreign companies that manufactured televisions. Bowe was only too happy to make this effort to gather more support, and while he was visiting East Asia, Sy DuBow, the legal director of the National Center for Law and the Deaf at Gallaudet University, led the effort to draft the legislation. Among the many people he collaborated with were John Ball, Larry R. Goldberg, Jeff Hutchins, Bob Richardson, and Karen Peltz Strauss.[62]

On November 21, 1989, Senator Harkin (D-IA), with lead cosponsors Senators John R. McCain (R-AZ), Daniel Inouye (D-HI), and Paul Simon (D-IL), took the long-awaited step and introduced S. 1974, nicknamed "the decoder chip bill."[63] Importantly, to provide even more rationale for the decoder chip, Sy DuBow had furnished Harkin with several literacy studies that revealed the benefits of captioning as an educational and motivational tool for people learning to read.[64] Harkin noted the ability of captioning to increase reading comprehension, language retention, and word recognition in deaf students.

On March 1, 1990, Congressman Major Owens (D-NY), along with Representatives David Bonior (D-MI), Edward Markey (D-MA), and Steve Gunderson (R-WI), announced their intent to introduce the House version of the decoder legislation. There was resistance from the Electronic Industries Association, which, like the FCC, as Strauss explains, claimed that it was "inherently unfair to force everyone to pay for decoder chips when these devices would benefit only a small minority."[65]

Then came hearings on Capitol Hill, where a lineup of celebrity witnesses testified. For the House subcommittee hearing held on May 2, 1990, this included such stars as Linda Bove, who portrayed the deaf librarian on *Sesame Street*. Only a few weeks later, a Senate subcommittee held its own hearings, again with a cast of Hollywood celebrities, this time including Marlee Matlin, a deaf Oscar winner for her role in *Children of a Lesser God*. Senators were so convinced of the need for legislation that they reported out the bill to the full Senate Committee on Commerce, Science and Transportation before the hearing was even completed—something outside their typical routine for moving bills.[66] By June, the Senate Commerce Committee sent the bill to the Senate floor, where it was passed on August 2, 1990.[67]

However, in order for a bill to become a law, it has to pass both the House and the Senate. During the summer of 1990, industry resistance began hurting the House bill's chances of ever getting to the House floor for a final vote. Because it had only been approved by a House subcommittee, it still had to move to the full House Commerce Committee before a floor vote could be held. The bill's supporters saw a need to expedite the process, and TDI urged its members to express their support to Congressman John Dingell (D-MI), chair of the House committee, to move the bill to the House floor. Dingell accomplished bringing the bill to a successful vote in his committee on September 25, 1990. Six days later, the House unanimously passed the legislation, clearing the path for the president to sign the bill into law.[68] But while the bill was waiting for the president's signature again, another form of resistance began to emerge. In *A New Civil Right*, attorney Karen Peltz Strauss described what she referred to as a "scathing attack" against the provisions of the decoder chip bill, which surfaced in the *Washington Times* on October 8, 1990. She described the editorial as condemning Congress for requiring consumers to "burn" $20 (the approximate cost of a chip) every time they purchased a TV. Fortunately, this complaint was ignored, and on October 15, 1990, the Television Decoder Circuitry Act of 1990 was signed into law. This law required all televisions manufactured or imported into America with screens thirteen inches or larger to be capable of displaying closed captions as of July 1, 1993.[69]

"We are very happy," Charles C. Estes, executive director of the NAD said. "This is one of the parts of life that deaf people have not been permitted access to until the last 10 years, and then only in a very limited way."[70]

A dream had come true. Deaf and hard of hearing people no longer had to buy captioning decoders. They would soon be able to watch TV through closed captions with much more facility. Although Mac Norwood lived long enough to see closed captions bring television access to deaf and hard of hearing people, he sadly did not live long enough to see President George H. W. Bush sign the Television Decoder Circuitry Act into law. Norwood passed away on March 22, 1989. Indeed, many deaf people were vouching for captions having had an impact on their education and welfare. As an illustration, in 1988 the Marlee Matlin stated, "It's hard to say where I'd be today without closed captioning, any more than [hearing people] could imagine on television sets without volume buttons. . . . [Closed captioning] is my lifeline to everything that is happening."[71] Mac well deserves the title "Father of Closed Captioning."

# 12

## Winds of Change: Bill Stark and a New Role for the National Association of the Deaf

*Students don't magically learn from a video just by staring at it. What you do before, during, and after the viewing will make video-watching a real learning experience.*

—Bill Stark, "A Video Decalogue for Pedagogues: A Crash Course for Teachers on How to Use Accessible Video-Based Media"

The first three decades of the captioned films program was a period of significant sociological, technological, and cultural change for deaf people in the United States. The parallel revolution in telephone access, also fueled by activism, combined with the captioned film movement to substantially decrease the isolation experienced by deaf people in society.

With regard to Deaf culture, the research of William Stokoe, Dorothy Casterline, and Carl Croneberg in the 1960s helped to verify ASL as a "true language," and the creative performances of the National Theatre of the Deaf brought further exposure to the beauty of visual language and led to a rapid growth in the number of people taking sign language classes. Television first began featuring deaf actors when Audree Norton, one of the founding members of the National Theatre of the Deaf, was hired to appear as a deaf character in a 1968 episode of the popular detective series *Mannix*. The 1980s was a breakthrough decade, with deaf actors increasingly making their presence known in the entertainment industry. Phyllis Frelich won a Tony Award for Best Actress in the play *Children of a Lesser God* in 1980. In 1987, Marlee Matlin won an Oscar for her performance in the 1986 film *Children of a Lesser God*, and the movie industry at last caught up with network television. Late as it was, this represented a significant change. Today many of the deaf characters that appear in films and television are played by deaf actors, enabling hearing audiences to see a wide variety of sign language usage, especially ASL, and bringing awareness and respect to Deaf culture.

In education, Norwood's program provided a great impetus for advancing the use of instructional media to new heights in K–12 and postsecondary education of deaf students, and the establishment of NTID at RIT, as well as other postsecondary programs around the country, led to an appreciable impact on the employment of deaf people in professional fields.

## ERNEST HAIRSTON: THE LAST "CHIEF"

These were a few of the exciting developments that had occurred in the United States when Ernest "Ernie" E. Hairston, a deaf Black man, became chief of the Education Department's Captioning and Adaptation branch in 1989. Born on March 1, 1939, in Stotesbury, West Virginia, a mining town, Hairston became deaf at the age of five from spinal meningitis. At first, his mother resisted the idea of sending him to the West Virginia School for the Colored Deaf and Blind near Charleston; she felt it was too far from home. Ultimately, however, she and young Ernie's father decided it was the best way to ensure a better life for their active son. During Hairston's sophomore year, in 1954, the US Supreme Court ruled such segregated schools unconstitutional. The Black students from his school were transferred to the West Virginia School for the Deaf and the Blind in Romney. Hairston viewed his experience in these schools as one of the best things to ever happen to him. Not only did he receive a solid elementary and secondary education, but the West Virginia School for the Colored Deaf and Blind was located a short distance from West Virginia State College. Many Black students in the college's strong teacher preparation program practiced in Hairston's classes and, although they were less familiar with deafness, they provided him with a challenging learning environment.

Hairston completed his bachelor's degree in education from Gallaudet College and held positions as a teacher at the Governor Morehead School for the Deaf in Raleigh, North Carolina; at the Michigan Association for Better Hearing and Speech in Lansing, Michigan; and as coordinator and supervising teacher at the State Technical Institute and Rehabilitation Center in Plainwell, Michigan. He also served as director of the Diagnostic, Evaluation, and Adjustment Facility in Columbus, Ohio, as he studied at Michigan State University. In 1967, he completed his master's degree in special educational administration at the California State University at Northridge's Leadership Training Program.

In 1971, Mac Norwood encouraged Hairston to apply for a position as an education program specialist with the Office of Special Education and Rehabilitative Services (OSERS). Through the years to follow, Hairston served in various capacities, and he published numerous reports on multimedia, career education, multicultural services, and captioning, as well as microcomputer, videodisc, and telecommunications technologies. By the time of Norwood's retirement in 1988, Hairston advanced through the ranks and was well prepared to take over Norwood's responsibilities within OSERS.

In addition, Hairston and Linwood Smith worked together to advocate the importance of addressing the problems associated with the lack of educational opportunities as well as underemployment, communication skills, and self-identity among Black deaf individuals. Their work together culminated in the publication of the book *Black and Deaf in America: Are We That Different?*[1]

## THE NATIONAL ASSOCIATION OF THE DEAF CAPTIONED FILMS ERA BEGINS

The passing of the ADA in 1990 was a milestone that greatly affected CFV. In addition to prohibiting discrimination against people with disabilities in employment, transportation, state and local government services, and businesses that are public accommodations or commercial facilities, it also required nationwide telecommunications relay services to allow people who are deaf, hard of hearing, or speech disabled to make calls over the telephone. Title III of the ADA required that public facilities, such as hospitals, bars, shopping centers, and museums (but, importantly, not movie theaters), provide access to verbal information on televisions, films, or slide shows. Although stiff opposition from the motion picture and cable industry prevented the ADA from including a requirement for closed captioning on television, as a concession

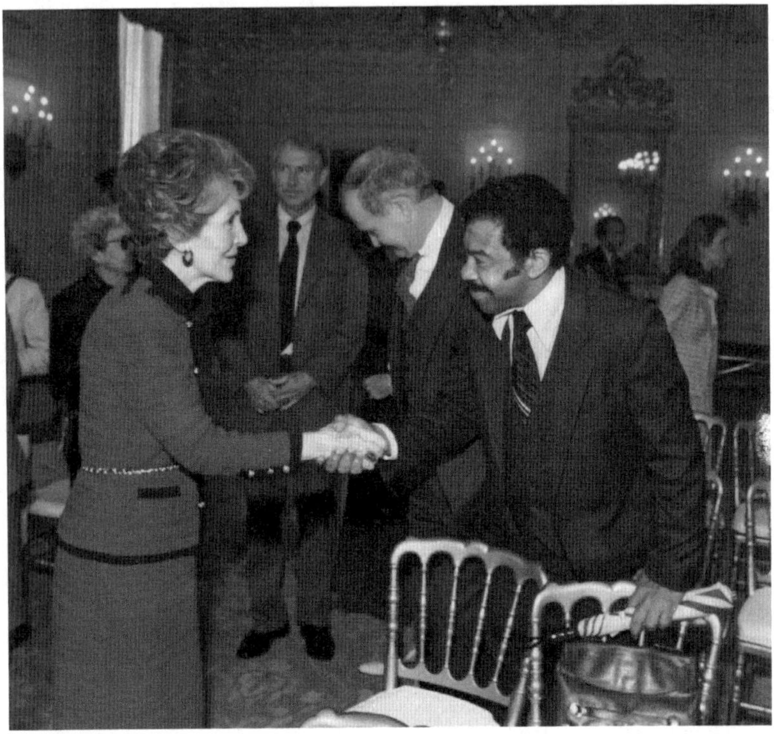

Ernest E. Hairston meets Nancy Reagan at a Recordings for the Blind Awards Ceremony. Image courtesy of Ernest E. Hairston.

Congress did include a requirement for all federally funded public service announcements to be captioned.[2]

This was a time when the winds of change were blowing. One catalyst for that change was the ADA, with its emphasis on the broad term *disabilities* and on full and equal opportunity for all citizens. Until 1990, deafness was the only disability addressed by the captioned films program. Also significant were the 1990 Amendments (PL 101-476) and the renaming of the 1975 Education for All Handicapped Children Act (EHA) legislation as IDEA. These changes focused on individuals with special needs and their preparation for the world of work and college, and affected the whole of the MSCF branch of the Department of Education. Among the provisions were added authorized training in instructional and assistive technology services aimed at dramatically increasing the number of technology projects. In serving a larger audience, captioning, in particular, gained ground in becoming a universal technology. Within a few years, the technology would be regulated by both the Television Decoder Circuitry Act of 1990 and the Telecommunication Act of 1996 and enforced by the US FCC.[3]

Both former and current CFV leaders were paying close attention to the potential impact of the new laws. IDEA improved on the EHA, providing a "free appropriate public education which emphasizes special education and related services designed to meet their unique needs," amended by PL 99-457 (EHA Amendments of 1986). Mac Norwood indicated two years before IDEA was passed that the "name of the game today is educational services" and that the role in the government's MSCF branch included research in technology related to "handicapped" children. Norwood viewed as a top priority the utilization of television as a means of bringing deaf people further into the mainstream of the general population.

Len Novick caught an important omission in the 1990 amendments: "I was astonished to discover that the underlying language supporting captioning contained reference to 'film' only, with no mention of captioned television, which by that time had existed as a service for almost a decade." As Novick worked with congressional staffers, what became PL 101-476 was edited to firmly establish authorization for federal support of captioned television.

—Len Novick, "Captioned Films for the Deaf: My Perspective," *Described and Captioned Media Program*, https://dcmp.org/learn/375-captioned-films-for-the-deaf-my-perspective

In 1991, the federal government again broke the captioning activity into two contracts: one for evaluation, selection, and captioning, and the other for distribution of films. The NAD became a major player in the history of captioning when, in October of that year, it received its first CFV contract award from the US Department of Education to administer the film and video evaluation, selection, and captioning. Modern Talking Picture Services (MTPS) retained the responsibility for distribution.[4] In addition to these changes, the funding award was no longer designated as a "contract"; it was henceforth a "grant/cooperative agreement."

Film distribution under MTPS involved two deaf leaders: Larry Stewart and Charles Estes. They traveled to South Carolina to convince Bill Stark to author the CFV grant proposal and take a lead in the project if funding was obtained. Working with NCI and former CFV project director Len Novick, who was on staff at NCI, Stark authored the proposal. After the grant was awarded, Stark became the new project director working out of a new satellite office in South Carolina. He worked efficiently with staff from Gallaudet University and used the services of NCI to caption educational films and videos in a shorter time frame than was possible through the annual summer workshops. In addition, an increasing emphasis on verbatim captioning was implemented, rather than caption writing with language adaptation for deaf children. The educators involved were committed to carefully monitoring the quality of captioning from the NAD-NCI collaboration to determine how the captions could best meet the educational needs of deaf school children. Captioned productions would continue to cover all levels of instruction between preschool and adult education, encompassing all major areas of the curriculum, and come with teachers' lesson guides that summarized the content of the films or videos, highlighted vocabulary, and suggested classroom follow-up activities.[5] It was one of the first of many changes made by Stark to develop CFV as the new captioning program under the NAD into a highly efficient operation.

Among those who supported the grant proposal was Bob Davila, then assistant secretary of education under President George H. W. Bush, and Ramon Rodriguez, the Education Department's CFV project officer. Whenever the Media Services and Captioned Films branch sponsored workshops or meetings or initiated new projects, Gil Delgado made sure that Bob Davila was among the participants.[6] Importantly, both Davila and Rodriguez were deaf and understood the importance of the project for both deaf children and their families. In addition, Nancy J. Bloch was on her way in as the NAD's first female chief executive officer. Davila, Rodriguez, and Bloch provided excellent guidance and support as Stark led CFV. Rodriguez was a friend and mentor to Stark and was passionate about CFV and the needs of the Deaf community for captioning.

As project director, Stark did not involve himself in the media services part of the branch, which was heavily focused on television and closed captioning. This meant he was able to focus exclusively on captioned films, carrying on the original intent of the 1958 legislation.[7]

Stark had been with CFD on an almost continuous basis since John A. Gough was leading the program in 1966. He also enthusiastically acted as a film and video evaluator, a captioner, a lesson guide writer, a summer lesson guide workshop director, a spokesperson, and a consultant. Mac Norwood taught Stark how to caption films, and Norwood became another mentor to him for all his activities. He placed great emphasis on the creation of lesson guides for the educational media activity.

As a tribute to his accomplishments in these early years, MTPS wrote about Bill Stark that he was " . . . an example to many of us of what we'd like to be, both personally and professionally."[8]

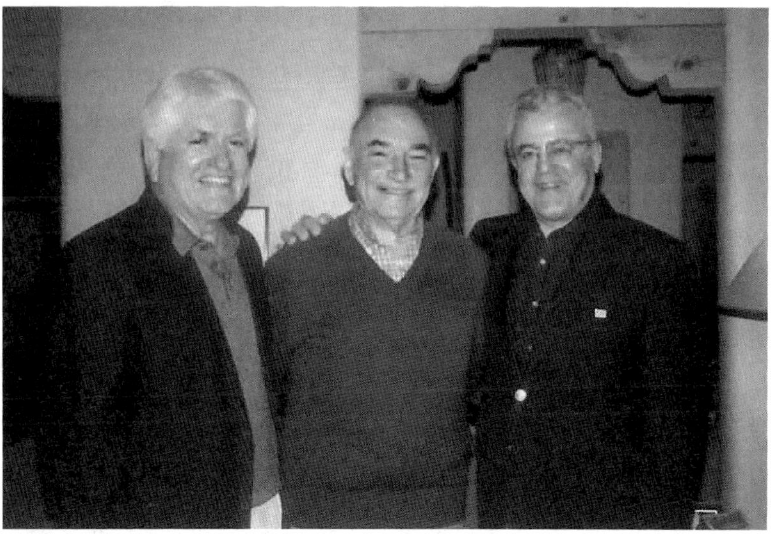

Captioning advocates (left to right): Ramon Rodriguez, Gil Delgado, and Bob Davila. All three Hispanic educators played critical roles in enhancing Deaf education through the use of media in the classroom. Image courtesy of Toni Atencio.

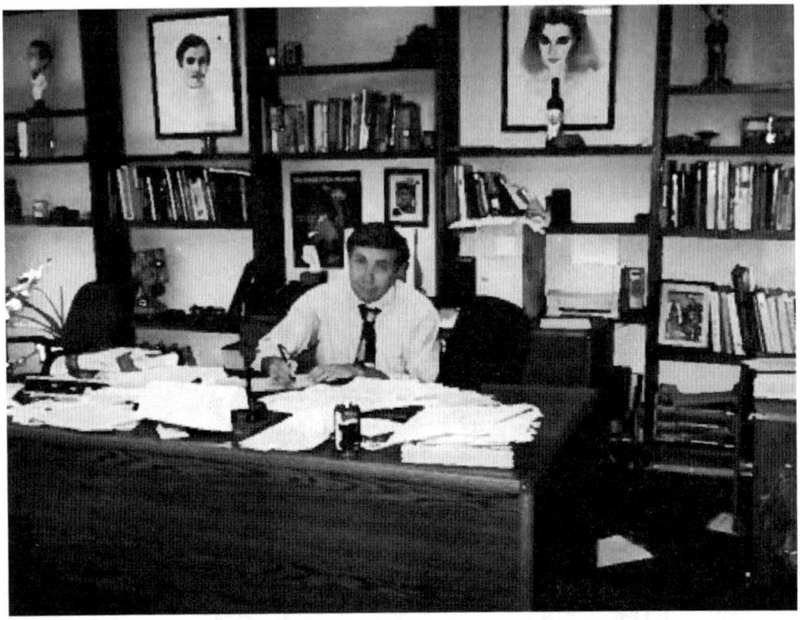

Bill Stark served as director of the CFV project from 1991 to 2011. Prior to assuming this leadership role, he had been with the project on a continuous basis since 1966. Image courtesy of Bill Stark.

Meanwhile, Ernie Hairston went on to serve as the agency expert and authority in the areas of educational media and media technology in the Office of Special Education Research to Practice Division. Judith Holt replaced Hairston as the government's contact for CFV. Hairston continued to participate and work with the Deaf community in many roles, and he provided valuable guidance to various institutions and organizations, including the NAD and the National Black Deaf Advocates (NBDA), of which he was a founding member and a driving force.

## CAPTIONED EDUCATIONAL FILMS

IDEA had a profound influence on those working to make information accessible to deaf students, including the NAD's captioning program. Now, with so many deaf and hard of hearing children being mainstreamed in public schools, Bill Stark and CFV would have to pioneer new research emphases to support these children. To accommodate the new emphases, during the first decade of Stark's leadership within the NAD, it was critical that he remain abreast with what was happening in the world of captioning.

As school programs adjusted to the new national legislation of the 1990s, some of the recommendations of the Ball State workshops and ideas generated in other early projects lingered in the minds of many educators, and a few schools continued to make use of the early products as well as the techniques and skills gained through workshops.[9] As a result of mainstreaming, however, many residential schools were dealing with declining enrollments, budget reduction issues, and meeting the educational needs of some newer and more challenging populations of students with severe and multiple disabilities. The early efforts of Gough's program to promote improvements in services to deaf students through the Ball State workshops also sparked interest in deafness by state-level special educators.

Importantly, in the early 1990s, the first few years of CFV operating in coordination with the NAD, there was a tidal wave of reform that has lasted until the present time. In the first full grant cycle, the evaluation of captioning vendors for the Department of Education and the posting of an approved listing of all vendors passing the evaluation marked a significant evolution in the captioning of educational films. Companies performing captioning work for the program in the NAD were required to pass this evaluation. When Dianne Stark, Bill's wife, and her team did not approve some vendors, the vendors took their complaints directly to the Department of Education. Each time there was a concern, the Department of Education consulted with the NAD and supported Dianne's response.

In addition, 1991 was also the year that the decision was made by CFV to stop making films available with edited captions. Edited captions were the result of the program weathering another challenge during its early years. An effort was made to edit the captions but not the soundtrack. Some teachers and hard of hearing students could hear the difference and found it annoying. To overcome that problem, the program began what they called "syncapping," or redoing the soundtrack to match the edited captions. Some teachers praised what they thought was the use of verbatim captions. But the program soon realized that, as Mac Norwood described in his paper titled "Just Don't Scramble the Wrong Egg," as educators it was important to deal with reading levels, not listening levels.[10]

The many research studies conducted before 1991 had also revealed additional controversial perspectives on whether films should include verbatim, near-verbatim, or edited captions. NTID, Gallaudet, television networks, and other users of captions recognized that audiences varied in age range, and preferences for the type of captions within each audience also varied. In 1991, CFV knew that within a few years, its primary audience would be K–12 deaf and hard of hearing students, but that did not necessarily make it easier to go with near-verbatim captions. Subsequently, some anger was generated, especially among people who felt strongly that the captions for deaf students should be edited. On the other hand, there were some who believed "dumbing down" the captions would lead to a lost opportunity for deaf film viewers to have their reading skills challenged progressively while learning content. Whereas some scholars often supported editing, Deaf associations tended to demand verbatim captions as the only way to have full access to audiovisual content.

With near-verbatim captions, CFV generally strove to meet a maximum words-per-minute rate based on grade level of the production. Although there was some editing, important vocabulary was maintained, and the language was not watered down. The goal was not to change the reading level of the captions but to reduce the number of words so that they could be read as the deaf student also interpreted the visual component of the video.

After 1991, the continued emphasis on near-verbatim captions, and the development of the *Captioning Key*, a set of guidelines and preferred techniques regarding how to caption educational media, eventually led to wider acceptance of this change. The *Captioning Key* is discussed in more detail in the next section.

Over the years to follow, each time a film was used, clients provided feedback on a form that collects qualitative and quantitative evaluation data. Most recently, in the current grant cycle, 99.4 percent of the user responses indicated that the captions were considered high quality.

Also, it was in 1991 that CFV again changed its name, this time dropping "for the Deaf" in deference to the hard of hearing community.

## THE *CAPTIONING KEY*

Although CFD, WGBH, and NCI had rules for captioning, they were generally closely guarded. Under Bill Stark, CFV developed the *Captioning Key: Guidelines and Preferred Techniques* for captioning service vendors performing work for the program. The document, distributed in 1994, provided an overview regarding how to caption. It included information on types, methods, and styles of captioning; presentation rate; text; language mechanics; and dealing with special considerations such as sound effects, music, and foreign language/dialect. Dianne Stark spearheaded the effort. She had used captioned films and videos for two decades in her classrooms, and for CFV she served as a captioning/lesson guide workshop administrator for several years.

The *Captioning Key* became a reference for captioning of both entertainment and educational media targeted to consumers at all levels (children through adults). The manual was distributed internationally and was translated into other languages such as German. The information was applicable to all others that provide captioning of all types of media at various levels. Thus, these guidelines were useful to media producers/distributors and others who were considering captioning their products or learning about captioning.

In addition, in 1994, the exclusive contract with NCI to perform all captioning work for CFV was ended. Contracts were henceforth made with multiple companies.

> In 1992, the National Television Systems Corporation developed captioning standards with service providers, FCC, and the Electronic Industries Alliance. Canada Captions, Inc., was formed for raising funds for closed captioning in Canada. Hillsborough County, Florida, and Fremont, California, became the first county and city, respectively, to caption in real time all government and school board meetings, funded by a surcharge on all cable TV bills. Cheetah released CAPtivator, a postproduction captioning system for prerecorded video such as movies and television programming.

In 1993, President Clinton's inauguration was the first live event to have both captioning and descriptive video service; it aired on PBS and was accessible for viewers with hearing or vision disabilities.

## THE MARYLAND CAMPAIGN

A chain of other important developments, both federally and locally, came about during this time, such as when the FCC adopted technical standards for closed captioning on cable systems in 1991. In 1992, while lobbying for real-time captioning of local news programming to the local networks to push for equality to information, Toby R. Silver, an independent producer and technical consultant, was preparing for the first Television For All (TVFA) meeting with NBC affiliate WRC-4 personnel in Washington, DC. Spearheaded by Silver, TVFA was a coalition consisting of more than twenty local, state, and national organizations. She produced an instructional video on the differences between live captioning and electronic newsroom captioning. Silver assumed the network personnel had no clue there were two types of captioning so she brought the video with her. At one meeting where WRC-4 staff were showing their newsroom captioning work to TVFA members, a live segment was introduced. At first, station personnel thought some technical problem was interfering with the captioning. The TVFA members at the meeting explained that what they were viewing was the result of the scripted captioning policy. These captions did not provide equal access to information. This experience was effective in leading WRC-4 to implement real-time captioning.[11] TVFA's success eventually led to all four local networks adopting real-time captioning of all local news programming, as well as WJZ-13 in Baltimore.

Two years later, on January 27, 1994, Silver and her friends in TVFA held a successful rally in Annapolis. They also provided testimony in support of HB 381, the Video Caption Access Bill, which would have required all videotapes sold in Maryland to be captioned.[12] Silver, along with Harvey Goodstein, Willis Mann, and Stan Schuchman, had met with lobbyists representing the Motion Picture Producers of America (MPAA). The MPAA representatives argued that they were doing a "good job" in closed captioning movie titles. "We, the deaf and hard of hearing consumers of captions," wrote Silver and her friends, "are here to tell you that they are not and that legislation is required to force them to provide captions." In briefly explaining the history of captioning to MPAA, the testimony mentioned how sixty years earlier, talkies were introduced. Silver wrote:

> Hollywood kicked us out of movie theaters throughout America. To make matters worse, the MPAA successfully lobbied the United States Congress for an exemption to the recently enacted Americans with Disabilities Act which stated that open captions were not necessary. If the MPAA is doing such a good job, why are they continuing to actively keep us out of movie theaters?[13]

Silver's testimony described the passage of the Television Decoder Circuitry Act of 1990 and how all television sets manufactured in America with a thirteen-inch or larger screen would have built-in captioning chips. Most videos in 1994, however, remained uncaptioned. "Even hugely popular TV miniseries like Alex Haley's *Roots* originally captioned for network television are not captioned in their recently released video format." "Mr. Chairman," she concluded, "I urge you and the members of the committee to pass House Bill 381 and make Maryland the first state to provide video access for all of its citizens."[14]

> WHY SHOULD WE WAIT FOR HOLLYWOOD TO BESTOW THEIR FAVORS ON US? IT IS TIME TO DISCARD HOLLYWOOD'S PLANTATION MENTALITY
>
> —From a rally flyer in 1994 during Maryland's Rally for a Video Caption Access Bill. Courtesy of Toby R. Silver.

Along with these Marylanders, Mike O'Donnell, Bob Weinstock, Alan Amann, and others were helping to organize the campaign to pass HB 381. Delegate Marty Madden, the chief sponsor, lined up forty cosponsors. The volunteers set up the rally, obtained speakers, lined up endorsements, lobbied delegates for support, and sent out daily email updates. Gallaudet students assisted with the preparation of the rally.

And people turned out on what had to be one of the coldest days of the year to attend the rally on January 27.

Unfortunately, due to pressure from the MPAA, this bill was defeated in the Maryland legislature. Despite this, deaf activists remained proud of their efforts. "While we did not win," summarized Willis Mann, "[the Maryland Association of the Deaf] fought hard . . . made a lot of progress . . . [and] received a lot of attention. . . . "[15]

When asked why she advocated so strongly for open captioning when the networks and studio heads, regulators, and legislators had been resisting for years, Toby Silver explained: "Captioning provides universal access and the cost is a drop in the distribution bucket, not a 'charity'; it benefits everyone, not just those who are deaf and hard of hearing, but it helps immigrants with limited English, increases literacy in children and assists the elderly; it should be a 'given.'"[16]

Silver and a host of other activists would also join the fight for captioning in movie theaters.

## THE PASSING OF JOHN GOUGH

After retiring in 1969, John Gough had spent much of his time traveling in Europe with his wife Harriett Nay Gough, gardening, and helping to publish language training and other deaf education materials through a company called Edutec. Harriett suffered through the painful effects of amyotrophic lateral sclerosis (ALS).

One hot July day in 1988, she was watching John replenish their birdbath in their beautiful garden when the heat and exertion caused him to have a heart attack. Harriett died on July 26 while John was in the hospital for bypass surgery.

Gough passed away on December 20, 1994. In the year before he died, this humble pioneer wrote about how proud he was of his family, grateful for the support they had given him. In his memoirs, he wrote about his many efforts to promote education for people of all races, gender, and age, in particular. This included, to name a few activities that reveal his heart, assisting a lawyer helping a family attempt to enroll the first deaf Black child in the Mississippi School for the Deaf, and even writing a letter to a bishop and remonstrating him, successfully, against the use of "Uncle Tom" jokes with an all-white church assembly. He also wrote that the captioned films story was the "most interesting and significant phase of my career."[17]

The year Gough passed, it had been thirty-five years since he had stood before NAD members in Dallas, Texas, and promised the American Deaf community that the government would enrich their lives through a government-sponsored program providing films with captions.[18] He lived long enough to see the NAD itself take a commanding role in the revolution he helped begin.

Subsequent leaders of the captioned films program were inspired by Gough's decade of dedication and pioneering work, especially Mac Norwood, Gil Delgado, and Bill Stark. Perhaps Gough's legacy is best summarized in a letter written to him by Frank B. Withrow, who had a deaf brother and had spent twenty-seven years in the Department of Education. Two years before Gough's death, in 1992, Withrow was serving as the team leader of the Technology Applications Group. He wrote to Gough in June about a national symposium of educational media technology at NTID in Rochester. Invited to be the keynote speaker on the last day of the conference, Withrow had studied the history of captioned films program from Gough's time until the 1990s. He told Gough: "I sincerely believe that one of the most significant things to happen in the betterment of all services for deaf people has been the caption program." He added that the government was putting very little into the overall captioning of TV, with much of this coming from the commercial market. And he saw Gough's work with captioned films and other media as "forerunners of what general education is doing today," especially the concepts that were still active and being used at the state and local levels. "The program of this national symposium reinforced my belief that the pioneering work you did still lives and is important in American education and especially education of the deaf. . . . I fully believe that the legacy of captions is the most positive thing that has happened in education of the deaf in this century."[19]

Gough had lived to see commercial television being closed captioned and caption decoders being built into every TV sold in the United States.

In 1994, the "Information Superhighway" speech by Vice President Al Gore became the first live event to be captioned in cyberspace. And in 1995, *Live! With Derek McGinty*, by Discovery Communications, became the first regular weekly show to be captioned on the internet. Kaleidoscope Television went fully captioned full time, including all programs, commercials, and anything with dialogue.

In 1995, the NAD was awarded the contract to administer the media distribution activity, taking over these responsibilities from MTPS. The NAD now held the contracts for both selection/captioning of films and for their distribution. Although the depositories of educational films remained at residential schools, the depositories for captioned movies and special interest films were successively moved to several locations. Approval of a proposal was also granted to allow the NAD to procure the rights to film titles, absorbing into CFV this activity long performed by the Department of Education contracts office. As a result, the budget for the activities was increased. For the years 1995 to 2001 the annual budgets were approximately $3.5M.[20]

Several important personnel changes also took place in 1995. Bill Stark's son, Jason Stark, became the information technology director for CFV, and Ramon Rodriguez served in the role of the Department of Education project officer for the NAD.

In 1996, CFV began offering a captioned master back to any company granting rights to a title, hoping to negotiate a better purchase price. The successful strategy followed for the next two decades helped the company meet the mandate of the law with a product ready for television broadcast and with a broader sales appeal.[21]

Many developments related to television and movie captioning occurred as well. The Telecommunications Act of 1996 added Section 713 to the Communications Act of 1934, requiring the FCC to prescribe rules and implementation schedules for closed captioning of television video programs.

Importantly, the Telecommunications Act of 1996 required that all new video programming must be accessible through captions (to take

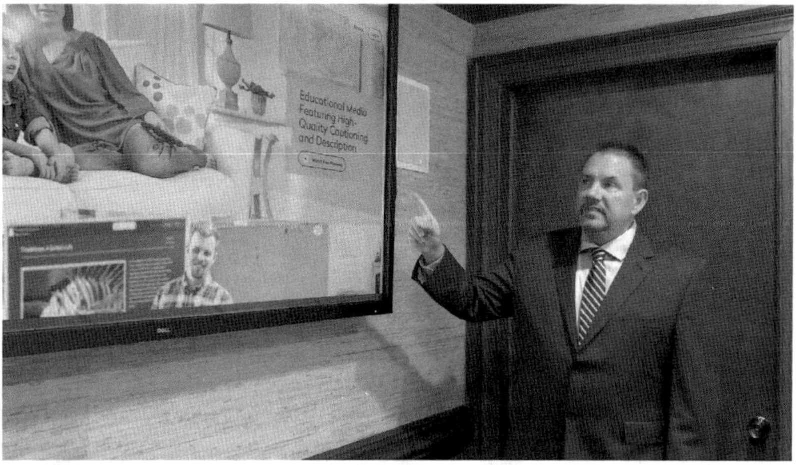


Jason Stark, who has been with the program since the NAD's first successful bid in 1991 to manage it, became DCMP program director in 2011. Previously, he performed as the DCMP information technology director. Image courtesy of the Described and Captioned Media Program.

effect in 1998). In compliance with the act's directive to establish rules for captioning, the FCC set up an eight-year timetable requiring an additional 25 percent of programming to be captioned in two-year benchmarks for English language programming. All captioning for new, nonexempt English language programs would be required by the start of 2006. Access for classroom videos was not within the FCC's jurisdiction and therefore was not covered by its rules, but many programs aired on TV would eventually find their way into the classroom. And because users of videos would be seeing more captioning, they would be encouraged to contact producers of videos to caption more.[22]

In addition, there was a requirement that all new Spanish programming had to be captioned by 2010; an extra two years was allowed for older Spanish programming. In her book, Strauss provides a schedule of captioning guidelines, including those applicable to both nonexempt and exempt programing. These exemptions still exist, covering nearly all commercials and even allowing new networks to forgo captioning completely for the first four years of their existence.[23]

The Society of Motion Pictures & Television Engineers formed a task force to develop captioning standards on MPEG and DVD formats the same year the Telecommunications Act was passed. Real-time reporters sent captions over the Internet Relay Chat channel and web page simultaneously. In 1997, the FCC adopted rules that gradually increased the amount of programming requiring closed captioning. Bill Clinton's second presidential inauguration in 1997 became the first inauguration to be simultaneously captioned live on television and the internet. In addition, Microsoft announced new Synchronized Accessible Multimedia Interchange computer and multimedia software captioning standards.[24]

Experiments conducted by the Silent Network, a national television network, resulted in the launching of programming in sign language, open captions, and full sound for deaf, hard of hearing, and hearing broadcast and cable viewers. The network began on local cable in 1983 and went national in 1984. It later grew to become a 24/7 network that was 100 percent captioned (all shows, commercials, and interstitials) on a voluntary basis before the FCC captioning regulations existed. It later evolved to become an OTT (Over the Top) television network streamed over the internet. One of Silent Network's deaf executives, David H. Pierce, operated the network's captioning department that accomplished its 100 percent captioning status. Pierce was responsible for more than 49,000 hours of national satellite broadcasts. In the latter half of the 1990s, Pierce served on the Closed Captioning Task Force of the National Cable Television Association, which provided input into the formulation of the television captioning regulations by the FCC. Pierce also worked with the Digital Cinema Initiative and the National Association of Theatre Owners on the specifications for captioning in digital cinema systems used in movie theaters.

# 13
## The Promise of Movie Theater Captioning

The problems that deaf and hard of hearing people experience with access to entertainment (theatrical or feature) films in theaters have been unceasing ever since the film industry began. As mentioned in chapter 1, when the silents were replaced by the talkies in the late 1920s, the notion of equal access for deaf and hearing people suddenly ended.

In the 1990s, Bill Stark and his staff at CFV were doing everything they could to fulfill the mandates from the Department of Education while attempting to meet the needs identified in the original intent of the 1958 legislation for entertainment films. For millions of deaf and hard of hearing people, viewing a first-run feature film in theaters remained problematic without adequate access through technology.

But what is at stake is not just an issue of having technology available to read the words on the screens. The issue of access is much broader.[1] Without true access, the goal of cultural awareness outlined in the earliest legislation for the government-sponsored captioning program would remain unfulfilled. The emerging technologies and growing activism around captioning in movie theaters is outlined in this chapter and ones to follow.

### REAR WINDOW CAPTIONING BRINGS HOPE

In 1993, with more than 750 hours of captioned shows being shown each week on television, including new releases, and cable channels just beginning to introduce captioning,[2] theaters remained largely inaccessible for the deaf population. That year, WGBH/The Caption Center devised the Rear Window captioning (RWC) system. The Rear Window technology was based largely on prior research and development conducted by the Corporation for Public Broadcasting and the WGBH National Center for Accessible Media in Boston.[3] This technology was codeveloped by WGBH and Rufus Butler Seder,[4] also of Boston. They also introduced relocatable roll-up captioning during the

Winter Olympic Games for CBS, which ensured that important action or graphics were not obscured with captions.

It was hoped that RWC, the first closed captioning system for theaters, would allow viewers to read closed captions without distracting others in the theater who aren't using the device that displays the captions. "We aimed to have a system that is closed . . . that would be available for every showing in equipped theaters from the day the movie opened," said Mary Watkins, outreach manager of Media Access at WGBH's National Center for Accessible Media. "To be able to see a movie when you want to see it with friends and families is a huge benefit of the closed system."[5]

Using acrylic panels that moviegoers could attach to the seat cup holders allowed the viewers to see reversed captions reflected from a light-emitting diode text display installed in the back of the theater. The captions appeared superimposed via the reflector panel on or beneath the movie screen. The panels themselves are portable, so users can sit anywhere in the theater.

In 1994, the RWC system was installed at the Langley IMAX theater at the National Air and Space Museum in Washington, DC. It was not installed in a first-run theater until 1996, at General Cinema in Sherman Oaks, California. Additional years were needed to expand the number to ten Rear Window–equipped theaters premiering throughout the United States.

The RWC system was costly, however, averaging about $11,000 to caption and describe each film. Many theater owners preferred the open captioning system, which does not require special equipment. In addition, the system was found to be disorienting to many deaf and hard of hearing viewers.

Indeed, deaf consumers often reported dissatisfaction with RWC. The red captions made for tough reading for some people with vision problems. Captions were sometimes blocked when another moviegoer stood up and blocked the reflected text or if a person passed by the deaf person's seat. The reflector panels did not stay in place, and choice of seating was limited because captions were viewed best from specific areas of the theater.[6] The battery may not have been charged.[7] In 2002, Katherine Bouton described it as "a clunky gooseneck screen that you carry to your seat and fit into the cup holder."[8]

## TRIPOD CAPTIONED FILMS

In addition to technology like the RWC system, there were organizations pushing for accessibility in theaters. Tripod Captioned Films (TCF) was the first to attempt to make open captioned, first-run 35 mm films available to local theaters. It was a nonprofit community outreach project of the TRIPOD (Toward Rehabilitation Involvement by Parents of the Deaf) Model School Program in Burbank, California, established in the early 1980s by Michael Shamberg and Megan Williams, parents of Jacob a deaf child. Through her Hollywood contacts, Williams planned an annual fundraiser with a movie premiere for the school program and, simultaneously, another theater would show an open captioned version to the Deaf community. But at that time, it was only about one movie per year. Cindy Costner, then married to actor Kevin Costner, offered to help, and she got Barbara and Chris Montan, also parents of a deaf son, Spencer, involved with Disney while Cindy Costner concentrated on Warner Brothers. Rather than working through the executives, they started with the filmmakers. There was a tremendous response to a film shown in Glendale, California. Within a short time, TCF had demonstrated a measure of success, and in 1993 the Montans turned it over to Nanci Linke-Ellis, the daughter of television producer Richard Linke. This became Nanci's life's work as an advocate for open captioned films.[9]

In time, TCF brought together other Hollywood studios to caption hundreds of films for deaf and hard of hearing Americans.[10] These included such hits as *Dances with Wolves*, *Titanic*, *Jurassic Park*, and *Star Wars: Episode 1—The Phantom Menace*. About fifty films were being distributed annually to 250-plus cities nationwide. Studios participating included Buena Vista Distribution, DreamWorks, MGM/UA, New Line Cinema, Paramount Pictures, Sony, 20th Century Fox, Universal, and Warner Brothers.[11]

However, beginning with only a single 35 mm print with captions laser etched onto the print, and without advertisements by the theater chain, these one-time "special showings" (until 1997) were usually not shown until months after the film's national release.[12] Linke-Ellis explained the hardships deaf people faced in gaining access to TCF. In the pre-internet era of 1993–1997, the deaf and hard of hearing patrons had such a lengthy wait to see a film print. Linke-Ellis also explained, however, that this predicament actually worked to their advantage:

> Given that they would have only one chance to see a film, it gave people advanced notice to arrange for baby sitters, take time off from work or put groups together for a social outing. If they missed the screening, they missed it. They would have to wait for it to come out on video (assuming captions were included, which most were not). This lead time led to widely attended shows, but on a very small, limited scale.[13]

During this time in open caption history, there were unique disappointments that occurred. For example, a special showing of the film *Schindler's List* was planned at a theater in Glendale, but due to a minor miscommunication, the print never arrived. Sadly, an elderly deaf couple had flown down from San Francisco to see the film. They had lived through World War II but had never shared the experience of watching a war film that was accessible through captions. TCF immediately rescheduled a future date, however, and the couple returned to view what would become a very emotional experience.[14]

On one hand, thanks to the exemplary efforts of Linke-Ellis and others at TRIPOD and TCF's successor organization, InSight Cinema, the studios had been willing to produce the open captioned films. Unfortunately, neither they nor the major theaters had ever made any effort to actually show these captioned films at times when the general (hearing) public wanted to see them. So, although Linke-Ellis pushed hard on this, and eventually was able to have some weekend showings, very few or no theaters were willing to have open captions on their screens around the clock.

## A VISIT TO THE STUDIOS

In 1996, Toby R. Silver and Karen Peltz Strauss took a trip on behalf of the NAD to visit all seven of the major motion picture studios to try to convince them of the merits of open captioned showings. They were accompanied by Nanci Linke-Ellis and Dana Mulvany, also staunch open captioning advocates, and were escorted by Van Stevenson of MPAA. Their goal was to convince the studio executives to provide more open captioned versions of their Hollywood films and to show these in mainstream theaters at times when people commonly went to the movies (as compared to the 11:00 a.m. open caption showings that were common). "While all of the studios that we visited generally acknowledged the need for access to their movies, and many, if not most, indicated a willingness to include closed captions on 'the digital prints of the future,'" Strauss wrote, "none were willing to commit to providing open captioning on films in movie theaters on a regular basis. Again, the concern was that not all hearing viewers might want captions on the screen while watching their movies."[15]

In 1996, the Movie Access Coalition (MAC), a predecessor of the Coalition for Movie Captioning, developed a white paper on the need for increased captioning of first-run movies in theaters.[16] MAC had been working with film studio executives and the National Association of Theater Owners to convince them to begin mainstreaming open captioned movies in area theaters on a weekly basis. The NAD's Toby R. Silver, chair of the MAC, helped lead these efforts. The MAC guidance included a proposed schedule to make all movies available with captions within three years. At a January 1997 meeting in California with representatives from the MAC, motion picture studios, National Association of Theatre Owners, and TCF, studios agreed to increase the number of open captioned movies. The studios also agreed to increase the number of prints available, show captioned movies at regular show times, and charge regular box office prices.[17]

> "The ADA guaranteed people with disabilities equal access to public accommodations, including theme parks. On January 18, 1997, in response to complaints brought against the Walt Disney Company by deaf consumers, the entertainment giant reached an agreement with the US Department of Justice to provide interpreters and other assistance for deaf and hard of hearing people, Disney agreed to incorporate sign language interpreters, innovative captioning systems, and other audiovisual aids into more than 100 rides, parades, attractions and shows at its four domestic theme parks."
>
> —Marla Dikerson, "Disney Agrees to Better Meet Needs of Deaf," *Los Angeles Times*, January 18, 1997.

## A NOT-SO-SIMPLE WISH—FORMAL TESTING OF AN OPEN CAPTIONED MOVIE

The persistence of the activists and organizations paid off, as movie executives eventually agreed to conduct a test project to distribute and show an open captioned movie in a limited number of cities. Previously, there had been other open captioned screenings of movies, usually donated by studios to nonprofit organizations acting as intermediaries between studios and theater chains. Often, these films did not reach theaters for many weeks or months after their noncaptioned versions, and they were often shown at off-hours.

The testing took place on July 20 and July 23, 1997, when four theaters in the United States were scheduled to show *A Simple Wish* starring Martin Short. These were the Cineplex Odeon theaters in New York City, Chicago, Universal City (Los Angeles), and Washington, DC. This was the first open captioned movie ever to have been booked from a major studio (Universal) directly to the theaters. With a large turnout, it was hoped that Universal and other studios would have greater incentive to provide other open captioned films in area theaters across the country on a weekly basis.

Behind the scenes there was added drama. Karen Peltz Strauss, one of the attorneys for the NAD at the time, recalled that when Universal Studios provided the open captioned version of *A Simple Wish* and the Cineplex Odeon Theaters agreed to show it, each wanted to see how many seats they could fill at mainstream times, such as weekends and evenings. But, she recalled, both the studio and the National Association of Theatre Owners refused to allow the Deaf community to publicize its showings in advance. This created a problem: MAC and the NAD knew they were headed for trouble if the theaters weren't filled.

On July 3, 1997, MAC went ahead anyway and distributed a press release several weeks before the showing. It was titled "First Open Captioned Movie Test Underway!!" MAC's message to the Deaf community was, "Show your support for open captioned films! Attend 'A Simple Wish,' enjoy a great movie, and demonstrate your interest in going to the movies!"[18]

During the third week of that month, newspapers ran reports with headlines such as "Audience Shares Rare Movie-Going Experience" and "Movie Subtitled for the Hearing Impaired." The Associated Press and other outlets' reports led readers to realize the importance of this breakthrough. "Many at the Sunday matinee of the new comedy-fantasy film, 'A Simple Wish' wondered if their first experience at simply buying tickets and easing into their seats like typical moviegoers would remain a rare experience or become routine." Silver was reported as saying that this was up to the film industry.[19]

However, trouble continued to brew. When a representative from the National Association of Theatre Owners found out that the press release was sent out without their permission, she called to scold Strauss. She was likely upset because of all the newspaper coverage the test film had received. It seemed to Strauss that the National Association of Theatre Owners and the studio wanted to sabotage the whole open captioning experiment. "Without publicity," she explained, "how were we supposed to fill the theater? I remember getting off the phone quite distraught—I was so angry and offended at how they were treating us."[20]

Then, of course, the theaters that showed the film were not all filled. There had been little to no publicity for some of the theaters and the movie itself was not critically acclaimed, as reflected in some of the reviews that followed the test.

Despite these efforts, there remained just a small number of captioned films being offered in theaters around the country. The success of open captioning would also create a problem for studios. For every first print, studios had to pay about $6,500 and about $1,500 for each additional print.[21]

In contrast with *A Simple Wish*, *Titanic*, which was released in late 1997, had such a good turnout of deaf moviegoers that it prompted Paramount to join the open captioned film program, providing eight

> For Patricia Johnson, open captioning provided a way for her deaf fifteen-year-old daughter to enjoy first-run films alongside her hearing friends and family members. "She doesn't want to be set apart from anyone else," said Johnson, director of operations for TCF.
>
> — Nanci Linke-Ellis, "Accessibility for Digital Cinema—A 2012 Update," *SMPTE Motion Imagining Journal* 121, no. 6 (September 2012): 90–94.

prints that were booked through the coordination of TCF. The prints were manually transported to 200 cities for a year.[22]

The demand for TCF's open captioned movies far exceeded the supply.[23] But the attendance at open captioned releases was limited to special screenings in select cities, and with only one or two showtimes daily across 450 cities, moviegoers had to vie for tickets to sold-out screenings before films moved on to the next city. "Showtimes are our biggest stumbling block," Patricia Johnson, director of operations for TCF, said. Movie studios were not legally required by the ADA to offer open captioned films. The studios had shown an interest in offering designated open captioned screenings, but were proceeding with caution. "They need to make sure there's an audience out there . . . We're proving that there is."[24]

## ACTIVISM GROWS

Activists advocating for open captions began to appear all over the country. They included large organizations such as the NAD, smaller groups such as local and state organizations serving deaf citizens, and individuals such as James "JJ" Jones and Carol Finkle; attorneys like Karen Peltz Strauss and John Waldo; and consumers who organized services to support local efforts to provide open captions, like Nanci Linke-Ellis. In this section, several of the individual efforts will be highlighted to illustrate the wide range of experiences among the activists.

One advocate who has spent an enormous amount of time and energy on access to first-run films in theaters is Carol Finkle, founder and director of the Philadelphia-based nonprofit Creative Access. This organization was founded to use the full range of the arts for social change. It hopes to lift the veil of near total invisibility of members of the nation's Deaf community by providing opportunities to participate across all genres available to hearing people. Finkle has been working since 1993 to bring open captioned first-run films, and local and Broadway touring companies, to deaf and hard of hearing audiences through ASL interpreting and captioned performances.

Finkle is a former teacher of deaf students. Her two children, Joshua and Abigail, are deaf. The family grew up in the shadow of Philadelphia's cultural center in the 1970s, craved movies and

Broadway performances—but never had the opportunity to enjoy them. "We were left out," said Finkle. So she turned her frustration into a crusade that has lasted decades.[25]

When interviewed in 1998, she was particularly excited about the first-run feature films that were coming out with open captions, along with every other first-run film waiting for release by Hollywood studios.

With her deaf family members and friends in mind, Finkle's crusade has focused on "the one thing that made us feel the worst—exclusion from the cultural life [of our city and region overall]."[26] In an endless series of battles, she has faced down Hollywood's bureaucracy, theater owners reluctant to give over a screen at a decent hour for a captioned movie that might drive away hearing patrons, and limited resources.[27] Her modest budget came from grants, donations, and public funds.

In 2001, *Harry Potter and the Sorcerer's Stone* was open captioned. One newspaper reporter wrote that it was a special experience, not only "because the blockbuster about the boy wizard has every kid in America heading for the movies" but because a teenager from northeast Philadelphia being interviewed said that for the first time in her life she actually understood every word uttered.[28] United Artists Riverview theater had shown the captioned screening of the movie to the delight of 350 deaf and hard of hearing students, many of whom sported round black glasses and lightning-bolt "scars," just like Harry. "It is extremely frustrating to try to enjoy a movie without captions," said a student from the Pennsylvania School for the Deaf, as she waited impatiently for the movie to start. "Captions are the only way of knowing what the characters say."[29] Three hours later, she was all smiles, signing excitedly about the movie.[30]

Finkle's eyes filled with tears over the students' expression of appreciation. Finkle, who was then celebrating a decade of advocacy for the roughly 450,000 people who are deaf or hard of hearing in the Philadelphia area, had helped local newspaper reporters see the broader goal of her work—bridging cultures. She emphasized that the world of arts is not just for hearing people.[31]

Harry Potter movies, of course, were popular around the country. The frustration of not being able to access them easily was also very common. Families were eager to see *Harry Potter and the Chamber of Secrets* when it began showing in area theaters, and they were disappointed when none of the dozen screens in some areas had a captioned version. These families appreciated captions, which made it possible for them to share the enjoyment of movies together. Barbara Raimondo, the mother of two deaf children, who was a government relations liaison, a director of advocacy, a parent consultant, and an attorney, was also frustrated. "Although there are lots of blockbuster movies out this time of year, we found nothing with captions in the movie pages during my children's two-week winter break from school," she said. "This means my kids don't get to go to the movies like hearing children and enjoy an afternoon eating popcorn and watching the latest flicks." Raimondo asked, "Why does Hollywood treat my children like second class citizens?"[32]

In Philadelphia in 2002, Carol Finkle organized a rally with about seventy-five people, including former Miss Deaf America Meghan Rainone and teen heartthrob Mark Webber. The rally was a protest over the lack of open captioned movies for deaf and hard of hearing moviegoers and their families everywhere. Often, Finkle would tell reporters, thousands find themselves needing to choose to either go to a movie and leave deaf relatives or friends at home, or not go at all. The protest was staged at the United Artists Riverview theater in Philadelphia. "They [moviemakers] don't believe there's a market here," Finkle told the crowd. "They're ignorant."[33]

Rainone was at the time of the rally twenty-three years old and in her third year at Temple University Law School. She told a reporter that she would like to see a film on a date but access was very difficult. "Can you imagine going to the movies and telling a guy, you have to interpret for me?" she asked a reporter.[34]

A second example of an individual activist for open captioned films is James "JJ" Jones, a deaf disability rights advocate for Whole Person, Inc. In 1996, he saw his first TCF-distributed open captioned film in Minneapolis at a dinner-theater event. He was stunned to see so many deaf people attending that he was forced to watch the film *Forest Gump* standing up. When he was about twelve years old, in 1966, his mother, a teacher of deaf children, brought home an old 16 mm projector to show him *Sergeant York* in open captions from CFD. He was thrilled to understand a film so easily. Then, while a college student at NTID, there were regularly offered open captioned films. After returning to Kansas, Jones would sometimes go to the local Deaf club to see old

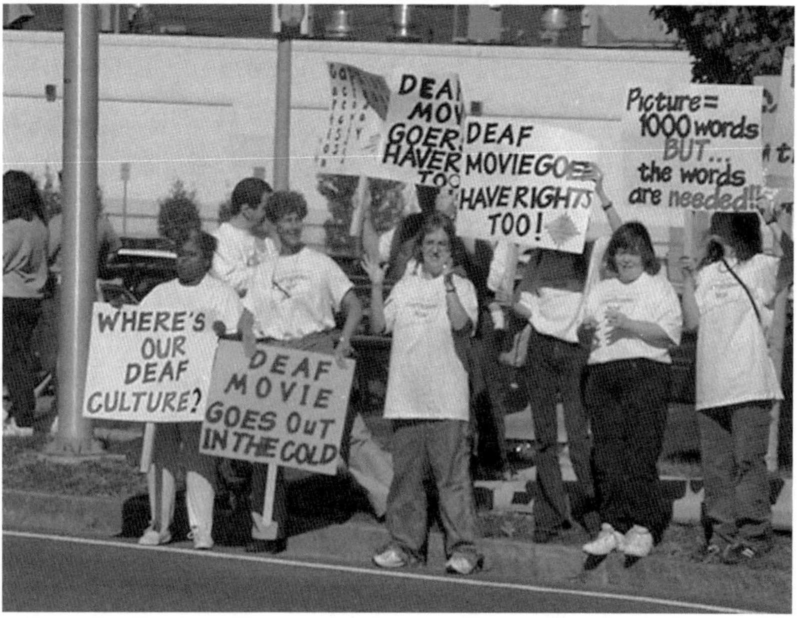


The Deaf community came out in droves to protest exclusion from first-run movies in theaters. Image courtesy of Carol Finkle.

16 mm movies, also from CFD. However, after the film in Minneapolis he realized that something bigger was needed for Deaf communities. That evening he asked about how the TCF movies were obtained and was told that he would need a nonprofit organization. He returned home and persuaded the Kansas Association for the Deaf and Johnson County Community College to help him plan for the first captioned film, *Mr. Holland's Opus*, to be exhibited at a theater at Metcalf South Shopping Center in April 1996. "We had to make our own tickets to sell for three months. More than 300 people bought advance tickets and another 100 bought tickets at the door." The theater manager didn't expect such a large turnout and had to show the film in his twin-theater facilities by passing the same film reels between the projectors. Popcorn ran out fifteen minutes before the movie started. Jones reflected, "I will NEVER forget the enthusiasm this event caused!"[35]

In 1998, Jones helped to get AMC Studio 30 in Olathe to show open captioned movies daily with three or four showings per day. This stopped several years later when AMC switched to RWC technology.

Then, when Jones and his friends celebrated their fifth anniversary of open captioned movies in 2001 at AMC's Studio 30 theater, TCF officers from Burbank, California, came. More than 300 patrons joined them. Five years later, they celebrated the tenth anniversary at Regal Theatres and, again, more than 300 people attended. After 2007, Regal Theatres showed open captioned movies daily for several years. When they switched to closed captioned films and deaf people had to use a pair of special glasses to access the captions, some deaf viewers lost interest.

The enthusiasm over open captions convinced Jones to continue planning such films with his group. In July 2013, more than 500 deaf and hard of hearing people (and some hearing friends) attended the first open captioned drive-in movie, *Monster University*, in Kansas City—on a Thursday night. For years, these films were shown monthly or semimonthly during the summers.

In Rochester, New York, Robert Menchel from NTID crusaded to bring open captioned films to theaters. By November 1999, he had succeeded in convincing the Regal Henrietta Theatre to become the fifth theater in the nation to feature daily screenings of open captioned films. This was welcome news to the thousands of profoundly deaf and hard of hearing residents in Rochester. Deaf people were coming to the film showings an hour early and staying an hour late, prompting the manager of Marketplace Cinema to turn the lights up so that the patrons could see each other's sign language.

The lack of progress in releasing mainstream films available with either open or closed captions led to the reorganization of the broader national Coalition for Movie Captioning (CMC). Charter members of the new coalition were major national organizations committed to addressing the theater access needs of America's 28 million consumers who are deaf, hard of hearing, and late-deafened.[36] CMC "believe[s] that individuals who are deaf, hard of hearing, and late-deafened should be able to attend any showing of any movie in any theater at any time with equal access through high-quality captioning." The CMC felt strongly that their representatives should be involved in developing the captioning standards that they would have to live with in the future, including their family and friends.[37]

Toby R. Silver reported that by late July 2000, only seventy different theaters throughout the country were presenting open captioned films at least once or twice per month. In the Baltimore–Washington metro area alone, there was not one dedicated screen. Ten years after the passage of the ADA, open and closed captioning was available regularly on less than .01 percent of movie screens.[38]

Later, three other open captioned movie advocates, Fred Korman, founder of DEAF (Deaf Entertainment Access Foundation) in the New York City, Long Island, and Westchester county areas of New York State; Erik Nordlof, leader of DC Deaf Moviegoers; and Jamie Berke, founder of Northern Virginia Open Captioned Movies, all witnessed and experienced the difficulties and frustrations related to the lack of open captioned movies, thus turning themselves into community organizers to bring attention to the need for improved movie theater access. They, too, are still actively working toward gaining a stronger foothold in the movement to remind movie theater owners/chains that this preferred captioning method needs to be moved to the forefront of reliable and cost-effective options for deaf and hard of hearing moviegoers. Engaging with movie theater owners/chains and their general managers has been their most prominent method of promoting the awareness and need for increased open captioned movie showings. They developed fact sheets that highlight their local deaf and hard of hearing population statistics, a detailed description of how open captioned movies work and the benefits derived from showing open captioned movies, and information on how to request open captioned movies through booking agents. In addition, the fact sheets explain effective ways to work with advocates in setting up and promoting open captioned movie showings.[39]

One unique program through DEAF is the Point of Contact ambassador program that assigns a deaf volunteer to be present at each of the open captioned movie theater showings to ensure that the captions are turned on and serve as liaison between the general manager and deaf/hard of hearing moviegoers if a problem arises. This allows for DEAF to be able to serve multiple theaters each week.

## THE STRUGGLE CONTINUES

Such activism was taking place in other towns and cities, but the expectations continued to fall short. By 2002 only a small number of theaters were showing first-run movies through RWC technology. Only six to twenty prints of any movie that was open captioned were made available, which meant about one out of every 1,000 screens in the United States was showing an open captioned film. The MPAA;s website listed 36,764 screens. The NAD reported that news accounts were saying that in 2001 TCF showed sixty different films in more than 500 theaters in the United States, but it was not explained that many of these theaters may have only had one showing each year or very sporadic showings throughout the year. Many auditoriums with RWC technology were not even using it.

Twelve years after the ADA was passed, voluntary encouragement had netted very low gains. Open captioned movies in theaters and numerous captioning technologies were not available when the ADA was passed. Informal surveys were undertaken by several national organizations. Consumers continued to feel strongly that being deaf or hard of hearing prevented them from seeing a movie. Information about captioned movie showings was very difficult to find. They believed that captioned movies should be available every day at all showings from the time they are released.

Despite the feeling of success when a film was shown with accessible captions, the constant struggle by individual advocates in cities and towns across the country had resulted in very slow progress nationally as well as at local levels.

## CAPTIONFISH

With the entertainment film distribution role dwindling due to the changing mandates from the Department of Education, open captioning still being in its nascent stage, and closed captioning technology experiments in theaters just beginning, the Deaf community was fortunate to have some entertainment through closed captions on television. However, both on television and in theaters the access to entertainment (and, therefore, the cultural awareness emphasized in PL 85-905) remained limited in the 1990s. The situation in theaters remained poor for years, especially in small towns. Volunteers worked hard to make individual theaters accessible, sometimes through the pressure of legal cases. Efforts to inform deaf people of accessibility were commendable. Captionfish, a web- and mobile-based search engine managed by

Linke-Ellis after TRIPOD turned over its program to InSight Cinema in 2002,[40] used the internet to inform deaf and hard of hearing theatergoers about which theaters were showing captioned films. Under the circumstances, Captionfish at least saved these viewers time in contacting individual theaters to see what was available and when.

Programs like InSight did not have efficient marketing at the time to reach many deaf moviegoers, and so while the number of captioned films increased due to growing demand, attendance for particular films fell.

More than a decade after the ADA was enacted in 1990, opening up doors everywhere for 53 million Americans with disabilities, many theaters were accessible to patrons using wheelchairs. Many of those same theaters, however, were not making their showings accessible to those with sensory disabilities such as deafness or blindness. "The cost to caption a feature film is a drop in the bucket compared with the total multimillion-dollar production budgets of recent blockbusters," said Cheryl Heppner, the deaf chair of the CMC in 2004.[41] Many believed theaters and the movie industry should address the potential audience that needs text to enjoy films.

Bob Menchel, a deaf physicist and educator, experienced frustrations while traveling because deaf people had to specially plan ahead to have captioning access. When he checked into a hotel room, he never found that setting up the captions was going to be the same as setting up in any other hotel where he had stayed. After unpacking, for example, in a hotel room in Manassas, Virginia, Bob decided to watch the news on TV. Looking at the remote, he did not see any dial that said CC, so he pushed the button for Setting. Nothing was there that was going to help him get the captions. Next, he tried Menu. There he saw Captions so he pressed that and a list came up that said English, Spanish, and Other. He decided to press English. That did not work so he kept looking. Finally, he found Closed Captions. Pressing that brought him to On/Off. He pressed On again and that turned the captions on. Now he had to go through another list: Text, Text 2, Text 3,

CC1, CC2, CC3, and so on. So he decided on CC1 because this seemed to work most of the time on other TVs. It now seemed that the system was set up and he could watch the news with captions. After watching the news, he decided to go out for dinner and he shut off the TV. After dinner he returned to his hotel room and turned on the TV to watch one of the programs. But for this television brand, the captions had disappeared and he had to go through the whole process again.

The next day he arrived at another hotel in North Carolina and checked in. Because it was time for the news, he looked at the remote and saw the button with CC on it. "This is going to be easy," Bob thought as he pressed the button and expected to see Captions On/Off. He pressed the remote and turned the captions to On, and again he saw CCI, CCII, CCIII, TEXT, TEXT2, and so forth. He pressed CCI and then waited. No good; he needed to start over. He pressed Exit. The captions came on and he could use the TV for the rest of the evening. He decided not to turn off the TV until he went to bed.

Waking up in the morning he turned on the TV to check the weather and news. NO CAPTIONS! He had to start all over again to set them up.

The next day, Bob was in another city and checked into a different hotel. Going to his room he again turned on the TV and looked at the remote to see how he could turn on the captions. Nothing there helped. He pressed Menu. Nothing worked. He pressed Settings. No help there. He looked behind the TV to see if there is some kind of button with CC on it. No luck. Finally, he went down to the desk and ask them to send someone up to turn on the captions. About half an hour later someone from the front desk comes in. He tried to turn on the captions and nothing worked so he called down to the front desk. Soon another person came in and she seemed to know what she was doing and Bob finally had captions on the TV. Next morning—he turned on the TV and again the captions had disappeared. Because he was checking out, he did not bother and just left the TV on.

—Robert Menchel, email to author, February 15, 2019

# 14

## New Horizons

Efforts to bring captioning to movie theaters were ongoing throughout the 1990s and 2000s, and during this time, Jason Stark and his technical team helped the NAD captioning program go online with its own website in 1997. His skills, particularly with the comprehensive website, made the program unlike anything in its history. With the creation of the website, clients were able to submit their film orders on the site and have films reserved immediately. The website became the heart of the program's operation.

Teachers also wrote comments about the benefits of media access. "I loved being able to provide my students with an accurate visual representation of *Romeo and Juliet* to support their English learning in the mainstream," wrote one instructor. "This was exactly what they needed to provide them with some background knowledge to aid in their reading and the video stayed true to the text." Teacher Cindy Malone explained, "Teaching Transcendentalism is hard enough for on-level students." She described the wide range of student skills in her class. One cartoon she used from the website, although simple, really helped to explain Thoreau's principles. She said it moved slowly enough that those who have to read the captions had plenty of time. Another teacher wrote about a student who was struggling with reading. "With the program materials [he] has now been able to understand and acknowledge the format and structure of sentences more. He has asked me a few questions about this, and has helped both of us understand each other's way of communication even more."[1] A fourth teacher expressed concern about how few social studies videos included closed captioning, even if they had been made or purchased recently. "Having access to [the program]," wrote Casey Zang, "has made a world of difference in my ability to create a classroom that is equal and accessible for everyone."[2]

Comments also came from families using the resources. "This program is helpful to communicate better with my deaf blind Grandson," wrote Teresa Ramirez.[3]

CMP logo. Image courtesy of Described and Captioned Media Program.

As accessibility information was being converted to digital format, CFV established a clearinghouse of information for how to caption, how to utilize captioned media in learning, sources of captioned movies and educational videos, and other information.

In 1998, CFV was renamed the Captioned Media Program (CMP) and stopped circulating films, focusing on videos and also adding CD-ROMs and other multimedia to its collection.

## ACCESS TO ENTERTAINMENT FILMS: A NEGLECTED ISSUE

Gaining access to feature entertainment films in movie theaters or neighborhood video stores remained problematic. In 2004 Jack Levesque, coordinator of the Deaf studies program at Holyoke Community College in Massachusetts, once wrote about the long struggle for access to entertainment films. "I can tell you," he explained, "we (Deaf) are a patient people because we have a history to prove it. Movies with sound were introduced in the late 1920s, and we had to wait 30 years for the first one to be captioned." After CFD began offering some captioned movies in the late 1950s, deaf people had to wait for a "movie night" at a residential school, a social club, or at a home gathering, which most often occurred about once per month. They were too few and far between.[4]

When deaf people attempted to find a film in a store, they had to practice much patience. They drove down to the local video store and browsed through several thousand videos to see which films were closed captioned. Some deaf people borrowed or purchased tapes only to return them because the captions were not working. "We had to wait another 30 years for captioned television . . . ," Jack Levesque wrote. And it was pure luck at the video store to find any hobby or travel or educational video that was available with captions. When VHS movies came out, deaf and hard of hearing people had to wait at least six months to see a captioned VHS rental.[5]

During the earlier years, CFD provided some films through distribution but it required a long wait time as well. Now, there still remained few good solutions to meet the high demand. Feature films containing captions generally remained unavailable for deaf audiences until long after they had been available to hearing audiences. This delay also occurred because purchase or lease of theatrical films by the US Department of Education and captioning of these materials was a lengthy process, often taking many months between the time a feature film first appeared in movie theaters around the country and the time that it was available for loan through CFV.

Some movies, of course, were never made available through CFV because the producer would not agree to captioning, the leasing/purchase costs were too high, or other reasons. Because of commercial restrictions on certain types of films, captioned feature films were restricted to use by deaf persons, and only groups that included three or more deaf persons were qualified to receive free loans. Borrowers were required to establish an account before being eligible to borrow captioned feature films.

## WRITING ON THE WALL

During the 1990s, the Department of Education began questioning the dual focus of captioning efforts on both education and entertainment. It was in March 1998 that Senators Joseph Lieberman (D-CT) and Dan Coats (R-IN) wrote a letter to the Department of Education expressing concern about federal funds being used to caption such programs as

*The Jerry Springer Show*. The indignant letter to the Education Secretary Richard Riley demanded that he cut off public money that helps caption such a "degrading and prurient program," which provided a "window . . . on the depths to which our culture has sunk."[6] Earlier in 1984, Lieberman also complained about the braille *Playboy* magazine in the Library of Congress. The censorship affected access, and it was just one of many issues the Deaf community faced.

When the NAD issued an action alert asking the Deaf community to write letters protesting the cutbacks in federal funding of entertainment films, many deaf people misunderstood and thought TV captions would be cut back. What the NAD did not mention was that the FCC was requiring that all TV shows be captioned. TDI sent an email clarifying that TV programs were not affected by this cutback. There was no danger of losing access to entertainment shows on TV because producers found private funding to keep those programs captioned.

A few months later, further criticism appeared in the *Congressional Record* on July 17, 1998, when Congressman Pete Hoekstra (R-MI) presented the Subcommittee on Oversight and Investigations of the Committee on Education and the Workforce report "Education at a Crossroads: What Works and What's Wasted in Education Today." He expressed concerns about the educational system being "fraught with bureaucracy and ineffective programs." The report presented a long list of federal education programs that purportedly did not focus on supporting what "works." It included the concern that "The federal government spends tax dollars on television shows like *Baywatch* and *Jerry Springer*. The Department of Education also funds closed captioning for other educational, programs such as *Ricki Lake* and *The Montel Williams Show*."[7]

This brouhaha presented some writing on the wall for Ernie Hairston and Bill Stark. And sure enough, they were informed that after 1998, CMP would no longer caption entertainment films or anything else for adults; the loan service to adults was also to stop at this point.

## A FIGHT FOR FEATURE FILMS

Bill Stark remained firm in his opinion that CMP should continue with entertainment films as well as educational films. In January 2000, he wrote a long letter to Debra Sturdivant in the Office of Special Education and Rehabilitative Services, US Department of Education, in response to her invitation to submit comments regarding what the terms "educational, news, and informational" mean in reference to the description and captioning of television, videos, or materials.

In his letter to Sturdivant, Stark stressed that the obvious goal of education is learning. And when most Americans hear the term *education*, they really think of instruction that most commonly occurs in a real classroom with real kids. But, he argued, learning occurs in a variety of situations and can be the result of either deliberate effort or unintended circumstance. "It often occurs without our consciously realizing it, such as learning verbatim two full verses of an obnoxious jingle of a television commercial or the theme song of situation comedy." Here, Stark mentioned television (including home videos and movies) as "the most pervasive provider of incidental learning. Without captioning, deaf children and adults are deprived of the opportunity for such learning."[8]

Stark's second important point was that although *entertainment* is most often defined as something that affords pleasure and holds our attention agreeably, "there is nothing in the definition that indicates that we don't or can't learn while being entertained."[9]

He stressed that there is considerable evidence that such learning does occur.

> For example, it has been said by both media theorists and historians that people's idea of history is formed more by novels, films and television dramas than it is by actual history texts. Who among us did not learn about issues of Black culture and history while watching *Amistad* or *Glory*? It was pleasurable for all of us to watch *Titanic* and *The Man in the Iron Mask*, but did we not learn a great deal from these historical novels brought to life? There are classic movie titles in the current . . . collection that are only available in captioned form to the deaf community because of the funding of the US Department of Education (ED). Many classic movie titles have never been captioned, and without ED support it is unlikely that they ever will be.[10]

His third argument to allow continued captioning of entertainment films was that literacy for all Americans has remained a priority of the Department of Education. But *literacy* has come to mean much more than the ability to read and write. It includes the ability to access, analyze, and critically evaluate communication in a variety of forms such as television, movies, and videos. Stark argued:

> Without captioning, the political, social, and economic implications and messages of these media are lost to the deaf community. For example, television is currently greatly influencing the process of our election of a new American leader, and it is crucial that candidates' debates and interviews be captioned. Further, television talk shows, situation comedies, dramas and other programming are also currently involving us in civil rights issues in Decatur (Illinois), the AIDS epidemic in Africa, turmoil at the highest political levels in Russia, health issues such as cochlear implants, and many others.[11]

And fourth, Stark wrote:

> Although it is true that no two Americans know exactly the same things, they often have a great deal of knowledge in common. This knowledge forms the basis for communities, and if it is shared by enough people, it becomes a distinguishing characteristic of our natural culture. To be *culturally literate* is to possess the information or knowledge needed to thrive in our society. The breadth of that information is great, extending over the major domains of human activity from sports to science. If they want to 'Be Like Mike' (Michael Jordan, basketball star) deaf children and adults must have access to the same network of information as other Americans. And this includes access to captioned sports programs. Is not "*Pokémon* both a Saturday morning cartoon and a cultural phenomenon? *Cultural literacy* constitutes the only sure avenue of opportunity for disadvantaged and disabled children, and the only reliable way of combating the social determinism that can condemn them to failure. The taxpayers of America do not benefit if expenditures for captioning are cut, but these expenditures are doubled later in the form of federal welfare assistance.

In conclusion, Stark encouraged the Department of Education:

> to broadly define the term 'educational, news, and informational' as it pertains to captioning. Home videos, Saturday morning cartoons, late night talk shows, Emmy winning television dramas, and other television programming provide indispensable reference points for the shared lives of Americans. Please continue to work toward helping 28,000,000 of our citizens acquire the knowledge and information that speakers, writers, employers, and the rest of society assumes they have.[12]

This was an excellent argument, but Stark was unable to convince the Department of Education. There was no response to his letter.

For a few months, Stark's program needed time to break down the video distribution mechanisms and to purge the registered account database of adults. The program's funding source was now the IDEA, and the emphasis of IDEA was on K–12 education. Consequently, the government-sponsored program had to completely drop captioned media services for deaf adults.[13]

As the twenty-first century began, the NAD was in a strong position to provide captioned educational materials to support deaf K–12 students. Its leadership also continued in the area of applying technologies to bring access to deaf people of all ages. In 2001, the NAD successfully bid to manage CMP for five additional years, giving the organization responsibility for both selection/captioning and for distribution of captioned educational films. That year, Ernie Hairston replaced Lynel McFadden as project officer.

## VIDEO STREAMING

In 2002, CMP began providing the country's first-ever streamed, open captioned educational videos through its website. Video streaming is the continuous delivery of a film through the internet to a remote user, allowing a video to be viewed online without being downloaded on a host computer or device.[14] CMP's move to video streaming was begun at this time due to a combination of being faced with a budget decrease and the wonders of the technology itself. More than 500 titles were available with English captions that year. Broadband supported up to seventy simultaneous users. Lesson guide production continued in full swing, with 528 guides being produced in 2001 and 2002, and there were 9 million hits to CMP's website. The NAD announced that "No other streamed material of this kind is captioned." They estimated that 28 million deaf and hard of hearing Americans are the potential beneficiaries of this historic media event. "Millions of Americans have been excluded from the internet's offerings of information, entertainment, and cultural content," said CMP's Stark. As the project director, he stated that the Department of Education–funded program had provided free-loan captioned media to deaf and hard of hearing persons, their families, and others who work with deaf individuals for over four decades. "This through-the-mail service will now be supplemented by the latest in digital technology."[15]

To receive streamed videos, consumers would need to have RealPlayer installed on their computers, which would enable them to receive both audio and video files. All videos streamed by CMP had open (visible) captions, with no additional software needed for viewing. The titles first streamed by CMP included productions such as *Blood, Heart, and Circulation*, *Giant Sea Turtles*, *Harriet Tubman and the Underground Railroad*, and *Lassie's Great Adventure*. There was no registration fee, and registered users could also borrow from CMP's collection of more than 4,000 open captioned videos.

Teachers across the country were indeed learning to enrich their classes through the use of media supplied by CMP. In 2004, dedicated instructors at the Kentucky School for the Deaf reinforced the importance of the streamed video resource. Vikki Rehberg wrote about how the deaf children in her class relied heavily on visual stimuli and sometimes had a limited experiential base. Brenda Prescott believed that captioned media provided a visual explanation of topics discussed in a class that often prepared the children for trips and hands-on learning. Evie Smith made use of the media in class and in small-group counseling sessions so that it could help show real-life situations and places that many students have never experienced. Jennifer Paycheck explained, "Captioned media help to take the children out of the classroom to places we cannot possibly visit." Jan Bradshaw wrote, "Captioned media. . . . provide a powerful connection between the printed word and meaning!" The lesson guides were helping teachers to expand their repertoire of instructional skills with the varied videos.[16]

## SPANISH LANGUAGE CAPTIONING

In the early 2000s, CMP's responsibilities were expanding, which included the captioning of an increasing number of films in Spanish. The captioning of films in Spanish began in 2001 and had reached 109 titles by 2003. Later in the decade, a reading incentive also focused on Spanish-speaking students and included a partnership with Dicapta, the first company to provide Spanish language video description and dubbing for blind and visually impaired students in the United States. Dicapta also participated in CMP's stakeholder panel on behalf of Hispanic children with disabilities. In return, partly a result of the collaboration and support from this company in promoting the availability of Spanish language media and translating CMP's *Captioning Key* into Spanish, Dicapta was granted funds from the US Department of Education under its television access program for five consecutive periods to serve the needs of Hispanic children with disabilities. The collaboration with CMP was invaluable.

John F. Levesque, was one educator who particularly liked the videos CMP offered that would "teach English as a second language to Deaf consumers who wish to improve their current skills." In expressing his appreciation, he also wrote about the extensive library of titles captioned in Spanish, and those that focused on special-interest topics. "For those lucky Deaf Americans who own or use computers," he wrote in 2004, "the [program] has free-loan captioned CD-ROM disks, and has begun video streaming of several hundred titles!"[17]

## DESCRIBED MEDIA FOR BLIND STUDENTS

In 2006, the US Department of Education announced a new priority—to incorporate into its program the technique known as "described media" for students who are blind or visually impaired. Description is the process of providing audio information about pictures, graphics, and other visual material that had been generally inaccessible to blind people in the past. Also known as "audio description," "video description," or "descriptive video information," the technological advance, as explained by the NCI, provides access to television and recorded video and film programming.[18] Beginning in 2007 all new educational media distributed by the program was expected to be not only captioned but also described. This has been a boon to blind people, who have been historically overlooked.

Example of a video description from the film *Frozen*. Note that this particular scene being described has no spoken dialogue.

". . . . a carrot nosed snowman shuffles up to a purple flower peeping out of deep snow. . . . he takes a deep sniff. . . . his nose lands on a frozen pond. . . . a reindeer looks up and pants like a dog . . . seeing the reindeer slip on the ice the snowman smiles and moves toward him though actually he's running on the spot the reindeer falls on its chin. . . . the snowman uses his arm as a crutch. . . . the reindeer paddles its front legs head over heels. . . . first snowman crawls along the ice. . . . the reindeer does the breast-stroke. . . . the snowman rolls his body but flips onto his back. . . . the reindeer's tongue sticks to the ice. . . . the snowman throws his head, twig arm and reindeer's lips tug [at] the carrot. . . . the carrot flies off and lands in soft snow. . . . the reindeer goes after it and snowman with his body parts hanging on his tail. . . . the Snow-man puts himself back together again and glumly contemplates his noseless state. . . . the reindeer jams the carrot back in place. . . . "

—"*Frozen* trailer with audio description," November 12, 2013, https://www.youtube.com/watch?v=O7j4_aP8dWA

After the script writers create descriptions of the program's visual elements, the script is time-coded and the audio or video cues are added to fit the descriptions. These are placed where there are natural pauses in the program dialogue. With more than 12 million blind and visually impaired American viewers or potential viewers, such access to news, information, and entertainment cannot be overstated.

Interestingly, although the first government-sponsored captioning program, CFD, was originally established to provide access for persons who were deaf or hard of hearing, the CFD pioneers Edmund B. Boatner and Clarence D. O'Connor, while seeking federal legislation in the late 1950s, examined the model of the existing Talking Books for the Blind program. In the early years of CFD, as depositories for educational captioned films were established at several residential schools providing services to both students who were deaf as well as those who were blind, some teachers of blind students found the auditory content of the captioned films helpful and used captioned films with both their deaf and blind students. Later, in 1992, the NAD, while managing CMP, conducted a feasibility study to examine the potential of a descriptive loan service for blind and visually impaired people. Results indicated strong support for such a service. Funding, however, was not available until 2006. The newly required service came as no surprise.[19] It had become inevitable that the program would be expanded to include this service under the Department of Education's Office of Special Education Programs, and funding was provided through IDEA.

With this mandate in 2006, the NAD consequently renamed the program again. This time, CMP became the Described and Captioned Media Program (DCMP) to reflect this broader focus of serving children who were blind or visually impaired in addition to the established role of captioning these same videos for those needing that access. DCMP immediately began collaborating with the American Foundation for the Blind to develop guidelines for adding descriptions to educational videos.

In an effort to level the playing field by making educational media accessible to all learners, Jason Stark and his technical team developed an online lending library to put valuable learning resources at the

DCMP logo. Image courtesy of the Described and Captioned Media Program.

fingertips of blind and visually impaired students and their families. Thanks to this resource DCMP began offering, they could learn about any subject that interested them—the human body, space travel, world history, and so on. Through these efforts, students of all abilities were given the opportunity to fully enjoy, instead of merely endure, films and other multimedia presentations.

The Department of Education was also responsible for television access projects for captioning and describing television programming, and the Video Description Research and Development Center (VDRDC), which investigated innovative technologies and techniques for making online video more accessible to blind and visually impaired students and consumers. JoAnn McCann was the project officer and served as a member of the VDRDC leadership network. Like Mac Norwood, McCann did much to expand television program options.

## CAPTIONING FOR PEOPLE WITH OTHER DISABILITIES

Over the years, people with disabilities other than deafness and blindness, such as those with learning disabilities, neurological disorders, or autism, have benefited from captions. Captions enriched the viewing experience for a wider audience. In one study conducted by Oregon State University, captions were found to be "very helpful" or "extremely helpful" to students with other disabilities, students with learning disabilities, and students who "always" or "often" had trouble maintaining focus.[20]

Captions appear to help students with a range of cognitive disabilities comprehend video-based instruction in a higher education context.[21] This includes students with autism spectrum disorder[22] and students with dyslexia.[23] The frontier of research on captioning is wide and exciting for all students. In general, the combination of audio and visual components in instruction may lead to the best learning environment possible. In one study titled "Presenting Effective Presentations with Visual Aids," for example, the US Department of Labor's Office of Training and Education tested research subjects' recall for information delivered orally, visually, or both orally and visually after a three-day time period. Those receiving oral-only instruction had an information retention rate of 10 percent, and those who learned through visual-only instruction had a retention rate of 35 percent. Importantly, the combination of both visual and oral instruction resulted in a 65 percent retention rate.[24] Such studies might lead to enhanced multimedia approaches.

## DCMP'S NETWORK

By 2006, when services for the blind and visually impaired were mandated, the newly named Described and Captioned Media Program included parents and officers from the American Society of Deaf Children (ASDC) and added the National Association of Parents of the Visually Impaired (NAPVI) and consumer group representatives like NBDA.[25] Other organizations involved with the DCMP board included the American Federation of the Blind, Dicapta Accessible Communication Developers, Council of Schools for the Blind, and the American Council of the Blind. William Daugherty, representing the Council of Schools for the Blind, viewed DCMP advisory group meetings as "impressive" gatherings of staff and members from both the Deaf and Blind communities. "It is apparent that both of these disability groups, consumers and professionals alike, hold [DCMP] in high regard due to its mission of providing access where access was once benignly denied."[26]

DCMP began tailoring each board member's role. For example, officers of the NAPVI and the ASDC routinely marketed DCMP to their organizations. Usually, everyone on the board evaluates media at some time or another. In general, the board members' roles have included oversight, marketing, advocacy, and media evaluation.

The NBDA served as an important member of DCMP's network. For more than three decades, the organization has been at the forefront of advocacy efforts for civil rights and equal access to education, employment, and social services on behalf of Black deaf and hard of hearing people in the United States. Founded in 1982, it has included more than thirty chapters across the country. The nonprofit organization is supported by its members and others interested in furthering its mission, vision, and strategic objectives.[27]

## THE *DESCRIPTION KEY*

In 2008, guidelines for adding description to videos were completed and posted on DCMP's website. The *Description Key* was the result of a two-year effort by the DCMP, the American Foundation for the Blind, and a select group of expert advisors. It contained instructions for choosing what to describe in a video and how to describe the video, along with an overview of technical issues and a list of resources that provide more information.

Description provides significant information for blind people and those with low vision, addressing their literacy and cognitive needs.[28] DCMP advises that the vocabulary and language structure used in a description be consistent with that used in the media being described. NAPVI has viewed DCMP's free-loan library service available to schools and families as imperative, and is satisfied that DCMP "has proven that they are the most qualified organization to develop advanced technology through mobile devices, streaming, and social networking features," building "a thoughtful broad selection of titles of high quality of described and captioned media."[29]

To effectively serve the blind and visually impaired people in its network, DCMP's work with the advisory board representative from NAPVI has led to quality control procedures for descriptions that involve checking each stage of the process, including the description script file, the edited project with added description, and the final mixed audio track master. The DCMP staff conduct a thorough review of the printed script of the description, and they check the grade level of the vocabulary and language, making sure that proper sentence structure is employed. The information in a description is presented in the present tense and active voice.[30] As a result, basic concepts presented in DCMP films are described with vocabulary that is meaningful to students. The content is described objectively in order to enhance student comprehension. Personal interpretation or censorship is avoided. Any sounds that are not readily identifiable must also be described, but information unavailable to the sighted viewer is not. The key encourages the sequence of activities to be described first, and additional details that aid in comprehension are provided as time allows. Scene changes in a film are described if relevant to unfolding action. Captions, including program titles, are read verbatim.

The second step of the DCMP description review process, which occurs after the description has been added, consists of previewing the project in the video editor software. Timing and rate of speech are checked for clarity, and the description must never be presented simultaneously with audio that is essential to comprehension of the content. Attention is especially given to the pronunciations of words, voice talents, style and delivery, and audio quality. Then, in a final examination there is a review of the audio file of the mixed program and description audio.[31]

With advice from its board members, DCMP developed the *Keys to Access* guidelines for both captioning and description to ensure that the accessibility features of its media were of the high quality required to provide equal access, following mandates by IDEA. The *Captioning Key*, captioning guidelines initially made available by the program years earlier, was completely revised for the web by DCMP staff. Both the *Captioning Key* and the *Description Key* were later revised to be compatible with mobile devices. Dicapta Accessible Communication Developers translated the *Captioning Key* into Spanish for DCMP. The keys are used extensively with Americans who are deaf or hard of hearing and those who have low vision or who are deafblind.

## CONSUMER SATISFACTION WITH DESCRIPTION

There is probably no better testimony to the value of description than is found in the letters written by users of the DCMP educational materials and their teachers and parents. Jason Stark has been collecting feedback for years from consumers of DCMP media. As an example of the hundreds of comments received by DCMP, Cecilia Robinson, a certified teacher from Houston, complimented the program for moving the instructional materials, learning resources, and assessments to an online platform:

> Most of the videos used in the classroom everyday are not captioned or described. As a result, many students with visual impairments, including those with deaf-blindness, are unable to benefit from the videos. Assistance from others become necessary. However, if videos are captioned and described, students will be empowered to use them independently. In my experience, students are happier and more confident when they know that they can do tasks on their own.[32]

"I am a blind author and lifelong learner who has benefited greatly from audio description. Audio description enriches my movie-watching experience because it allows me to pick up on subtle nonverbal cues, such as someone making a quiet exit, a passionate embrace, or an intricate dance sequence. Such plot-altering cues are lost when audio is your only way of taking in information.

During my school days, I dreaded films. I willed the time to pass quickly as a monotone voice droned on about coral reefs or African elephants. Most educational films, especially in Science class, communicated their messages through awe-inspiring pictures—tiny microbes under a microscope, a bald eagle gliding across the sky, or a cheetah running across the African Sahara. Those images were lost on me. The music playing behind them may have provided a nice accompaniment to my wandering mind but no real learning took place. If audio description would have been offered I could have gained a grasp of the size and scope of the inhabitants in that coral reef, marveled at the speed of the cheetah, or gained a deeper appreciation for our national bird."

—Mary Sefzi, general manager and owner of Audio Eyes, LLC, letter to Jason Stark, February 4, 2016

As Jason Stark established the impressive DCMP network, he also collected comments from organizations and company leaders who testified to the effectiveness of the NAD's program.[33] Bridge Multimedia, for example, dedicated to the creation and promotion of universally accessible multimedia, recognized that DCMP's on-demand archiving services are critical and has encouraged content providers wishing to receive video description services through its US Department of Education television access grant to consider inclusion of its materials in the DCMP archive.

The many different environments reached through DCMP also include ASL teachers who have found the resources provided by DCMP to be an invaluable educational tool. Ignacio Ponce, chair of the School of Modern Languages at Santa Barbara City College, wrote, "Many of

In 2005, the pioneers of closed captioning honored by the Accessible Media Industry Coalition (AMIC) during its annual Pioneers Gala included Malcolm Norwood, as "The Father of Closed Captioning." Jeff Hutchins, AMIC's executive director, stated, "Mac almost single-handedly launched a three-prong attack, funding PBS to develop the technology, funding WGBH to develop the operations, and lobbying network executives to get behind this effort."[36]

AMIC also honored some of the pioneers in real-time captioning. Martin H. Block was credited with writing the first "live" captions at NCI in 1982. In addition, Joe Karlovits, Marc Okrand, Betty Hallman, and Dave Margolis were recognized for their contributions to an industry that has had a positive effect on society. Judith H. Brentano summarized that studies at the Institute for Disabilities Research and Training show that the generation of young deaf people who have now grown up having captioning their entire lives are able to read better and faster than previous generations. "As a result, we now see more young deaf people entering professions that two decades ago were nearly devoid of people who are deaf, such as lawyers, doctors, and even politicians. Their access to mainstream media is credited as one of the major reasons for this societal shift that is helping tens of thousands of deaf people to live and work more productively."[37]

To this remarkable accomplishment, credit should be expanded to include all of those involved for decades in working in various projects with the CFP program (now the DCMP). As shown in this history, a great focus has always been placed on language skills development of deaf children in K–12 programs.

my students are themselves Deaf or hard of hearing, who were raised in the oral method and are beginning to learn ASL. The captioned and descriptive videos provided me with the opportunity to expose students to the rich history of Deaf and hard of hearing people in a way that is fully accessible."[34]

DCMP has also proved helpful for home schooling with both English-speaking and Spanish-speaking parents. Lucia Hernandez explained that her family uses DCMP in several ways to help her deaf son:

> I use this service to learn more about his native language [Spanish]. As an interpreter, I use your program to hone and perfect my skill set. My Deaf son and his hearing siblings use DCMP as an equal access tool to better connect with one another. This service is of vital importance to our family. I'm very happy in being part of this program. Thanks to the free on-demand described and captioned educational media, I've learned a lot of new signs. I can now communicate better with my son.[35]

## BILL STARK'S LAST BATTLE TO SAVE ENTERTAINMENT FILMS

In 2005, in addition to new priorities for improving educational films for blind and visually impaired students as well as for Spanish-speaking students, the winds of change included a mandate to completely phase out captioning and distribution of entertainment films. Bill Stark knew that the 1997 IDEA legislation had specified that all materials were to be educational, news, or informational. At this time, Lynel McFadden was project officer for the NAD grant award.

The deaf writer Max Friedman's prescient words in 1960 had come true. He worried that if deaf adults were not alert, they might discover one day that some pressure group had worked on the officers of the captioning project to change priorities to suit particular interests. He wrote that such was not the intent of the bill setting up the project, nor of those then in charge, but government bureaus are often responsive to pressures. "This is a call for alertness on the part of the deaf."[36]

In the opinion of the Department of Education, some of the films in the inventory did not fit into one of the three groups (educational, news, or informational), and the list of films had to be pared down further. In early November 2005, as he had argued in 2000 with Debra Sturdivant, Stark attempted to justify keeping some "theatrical" (entertainment) movies in the film collection. He knew that access to films in theaters remained poor, despite all the efforts with open and closed captioned film technologies. He received a response from Judith Holt, who had taken Hairston's place when Hairston moved back to the project officer post. He wrote, repeating some of the argument from his letter in 2000 to Debra Sturdivant: "It has been made very clear to us that movies (even very good ones) are not considered 'informational.' However, we have been making an exception for those that are based on literature."[37] Programs funded by the Department of Education, he emphasized, would be required to adhere to the new 2004 guidelines, which limit videos and so forth to those for classroom use.

Stark sought further clarification from the federal office that oversaw CMP. The classic movies that remained in the film collection were first released as entertainment features. "However," he explained, "the NAD has not retained such titles for entertainment purposes but for their educational and informational value." He wrote:

> Teachers continued to ask for "movie" titles that are invaluable and irreplaceable resources. Many of the titles had never appeared in any captioned form at all, and/or they were nearly impossible for people to acquire outside CMP. The overwhelming majority of movies that remained in the collection were literature based, though most have multiple educational/informational uses.[38]

Stark argued that the three titles Ernie used as examples in his email were based on literature. One was *2001: A Space Odyssey*. He wrote that this title was considered to be of sound educational use. The film was not only discussed in literature classes but also in teaching space, science, and technology, including about issues associated with the potentially destructive power of technological innovation in the military arena.

Similarly, Stark believed that *The African Queen* served more purposes than literature, as did *Beat the Devil*. The latter "is one of the few movies released in America of that era (1955) that was based on a novel written by a communist (also an Irish journalist). While this production 'entertains,' it also teaches about the 'corruptions of empire.'" Again, Stark stressed that this movie was unavailable to deaf users from any other place. It was classified as "adult" and was utilized at the teacher's/user's discretion.[39]

Some films in the collection for educational/informational use were not literature-based but were nevertheless "wonderful teaching/informational aids. They included *Abraham Lincoln* (covering the entire life of Lincoln), *The Big Lift* (reenactment of the Berlin airlift of 1948), *Children of the Struggle* (based on real events that took place during the Civil Rights struggle), *A Day in October* (based on facts about the movement of Danish Jews out of Denmark), and *The Jackie Robinson Story* (biography of Jackie Robinson starring Mr. Robinson).

Stark stressed that selected classic movies were valuable in teaching because of their ability to enhance appreciation of and understanding of literature, are great reading incentive material, generate enthusiasm for writing, and aid language development. They open up and revitalize studying science and mathematics, demonstrate the scientific method, and capture the rich details of distant times and places, major events of history, great achievements of civilization, and extraordinary men and women who have shaped the world. He summarized:

> We empathize with the difficulty of the task that the U.S. Department of Education has in interpreting and administering the reauthorization. We are further encouraged by the emphasis of Secretary Spellings on policy designed to help states better assist students with disabilities, and her pledge to continue working with states to ensure they have the flexibility needed to raise student achievement. She has emphasized a "common sense" approach to interpretation of No Child Left Behind and IDEA. We believe that the "movie" titles in our collection are educational/informational and should remain. Thank you for your consideration. I await your reply and further instruction.[40]

It seemed that the concerns expressed by deaf people four decades earlier were being realized. After CFD was first funded, there was a conference in 1960 in which John Gough had welcomed the opinions of deaf citizens. Notable among them was Byron B. Burnes, the deaf president of the NAD, whose words also rang out to the planners:

> I have said that the adult deaf favor development of the educational aspects of Captioned Films and we stand by that point of view. But may I emphasize again that providing films of a recreational kind is important, too. We who are deaf have limited opportunities for worthwhile recreation. Many of the things which hearing people take for granted such as radio, television, plays, and music cannot be enjoyed to any extent by the deaf. Captioned Films can fill a great need in our lives. We hope that this need will not be forgotten as the program pushes ahead.[41]

Bill Stark was advocating for deaf children what these deaf citizens expressed from the very beginning. Although deaf children thought some films were funny, some teachers intimated that such films should be "more instructive." But, unfortunately, his pleas to keep such films were lost in a sea of changes occurring at the time. In addition to having to stop distribution of entertainment films, CMP experienced a 60 percent budget cut, the national network of depositories for captioned films was eliminated, and the distribution of VHS videos was dropped. Also lost in this transition were "special interest" ("how-to") films that had been categorized as "entertainment." Many deaf people had become euphoric about closed captioned television and the beginnings of closed captions on rental movies. They felt it would all work out in the end, but they were, perhaps, overly optimistic.

# 15

## The Battles for Access Continue

*We can only achieve equality if all of you advocate for your rights.*
—Howard Rosenblum, National Association of the Deaf

Throughout the past few decades, the NAD, representing Deaf consumers, became increasingly involved in the battles over access to films. Local and state efforts to bring open captions to theaters were springing up across the country. The long battle for open captioned movie access in public theaters has taken many forms, but progress remained slow. The best thing deaf consumers could do was to support the NAD in its efforts.

Due to the resistance from movie theaters and the lack of equal accessibility, open caption advocates began working with localities and cities in various states to draft legislative bills with their local council members in hopes of having theaters provide open caption accessibility to deaf and hard of hearing movie patrons.

In 2008, members of the NAD at the national conference in New Orleans voted on the top thirty priority proposals to be sent to the Council of Representatives (delegate assembly). The delegates then selected the top twenty priorities of the NAD for 2008–2010. "Captioning in Movie Theaters" was ranked first: "The NAD shall advocate for all first-run movies to be shown with captions, preferably on-screen (open) captions, in all movie theaters across the nation."[1]

John Waldo, a deaf attorney whose practice focuses on legal issues involving deafness, compiled a concise four-part description of the history of some of the most important court cases involving open captions in theaters. When the Arizona attorney general's department sued the Harkins Theatres chain in 2010 for not providing captions and audio description for its movies, a federal district court judge in Arizona ruled that movie theaters are not required to display captions because captions alter a movie by changing audio content into visual content. Harkins won the first round. According to writer Howard Fischer, Judge Roslyn Silver said that "the nature of the [captioning] service means the disabled will never be able to fully and equally enjoy movies the same as others."[2] "The perfectly awful Arizona district court case," Waldo wrote, "went up on appeal. Since that was the first of the movie-captioning cases to reach the appellate level, we knew it would be really important."[3] Waldo and John Stanton also wrote an amicus brief.

Waldo explained that Judge Kozinski said that the primary challenge was accommodating the needs of deaf and hard of hearing moviegoers without altering the experience for everyone else, and this suggested that "open captioning was going to be a tough sell."[4] Although the ADA does not require open captioning, it did require closed captioning up to the point that providing captioning would be an "undue financial burden." In response to the Harkins Theatres argument, the court said that open captioning, "visible to everyone, could alter the movie-going experience for others, and could therefore be construed as a fundamental alteration."[5]

Meanwhile, the Washington State Communications Access Project filed a case under Washington State law against the corporate theater owners in the Seattle area. Waldo explained that the Deaf and Hard of Hearing Bar Association thought that if Harkins went badly, "we still had a shot under Washington State law, which is considerably more protective of people with disabilities than is the ADA."[6]

But the Washington trial court handed down a decision similar to the Harkins decision, and that outcome was affirmed on appeal by the Washington State Court of Appeals. Then, in July 2010, the Department of Justice (DOJ) proposed regulations that would require theaters to provide captioning in 50 percent of their auditoriums. DOJ followed up a Statement of Position opposing the stay Harkins requested pending final DOJ regulations. The DOJ said that accessibility is required by ADA, and that implementation should not await regulations. This was assurance that closed captioning would be available. Open captioning was not assured, and the question of "undue burden" would remain an issue for resolution in court.[7]

In 2011, AMC Theatres in Washington State were not going beyond what the DOJ required. The emphasis remained on the "undue burden" issue where the cost of providing caption-viewing devices should not exceed the theaters' revenues from use of those devices. But, as Waldo explained, AMC lost that argument and subsequently agreed to 100 percent compliance nationwide. In 2014, DOJ issued new proposed regulations, changing from the 50 percent requirement to calling for closed captioning in every auditorium of every theater (once that theater converts to digital projection).[8] Then, on December 2, 2016, final regulations were adopted requiring theaters to provide movie-by-movie notice that captioning is available, and on January 17, 2017, a final rule amended the DOJ regulation implementing Title III of the ADA of 1990. This rule requires that movie theater auditoriums provide closed captioning and audio description when showing a digital movie with those features "unless doing so would result in an undue burden or a fundamental alteration."[9] Movie theaters are required to have a specified number of captioning devices and audio description devices based on how many digital movies are shown in the auditoriums in a movie theater. The rule does not impose requirements for auditoriums that only exhibit analog movies.[10]

On April 25, 2018, Waldo wrote that many of his colleagues remained disappointed that the DOJ continued to refuse to require open captioning, "which most of us prefer." He asked, "Is there any hope there?"[11] In the last section of his series, posted on May 2, 2018, and titled "Movie Theater Captioning, Part 4 of 4: We're Not Done Yet," he summarized the state of open captioning in theaters as follows:

> While we like open captions, the theaters don't—or at least, they haven't. Their concern is that a lot of hearing people find open captions distracting. And they have some old data to back this up—prior to digital conversion, Regal showed some open-captioned movies at off hours, and attendance at the captioned showings was lower than attendance at non-captioned showings.
>
> My offhand observation is that people actually exposed to open captioning don't find it all that offensive, but a lot of people think they won't like it. From the theaters' point of view, that's actually the worst possible situation, because those folks won't give open-captioned movies a try. And from our point of view, the problem is that a number of the people who think open captioning would be a distraction wear judicial robes or are federal regulators. So even when we win smashing victories for closed captioning, we lose on open captioning.
>
> And yet . . . there is hope. I've always taken the position that lower attendance at individual open-captioned showings doesn't mean much, and that the real issue is whether some open-captioned showings reduce overall theater revenue. My hypothesis has always been that if open-captioned

showings are publicized in advance, people who don't want open captions will go to a different movie at the multiplex or go to a different showing, in which case the theaters don't lose any money. Conversely, people who really do want and need open captioning, even if relatively few in number, represent revenue that the theaters would otherwise not realize.[12]

Meanwhile, CMP had been at the forefront in providing accessible educational films. On October 8, 2010, President Barack Obama signed the 21st Century Communications and Video Accessibility Act (CVAA), PL 111-260, as amended by PL 111-265, which mandated a limited amount of video description on television programming. The FCC implemented this mandate by requiring video description on approximately seven hours of television programming a week on ABC, CBS, Fox, and NBC station affiliates in the top sixty markets in the United States. Fifty hours must be prime time or children's programming, and 37.5 may be any type of programming between 6 a.m. and midnight.

## ARE OPEN CAPTIONS DISTRACTING?

Numerous reports have come out about the benefits of captions. In one 2006 study of subtitle use by people who are deaf or hard of hearing conducted by England's Office of Communications (Ofcom), the regulatory body for UK television broadcasting, many television viewers reported benefits from subtitles even if they are not hard of hearing or deaf ("subtitles" in this study are the equivalent of closed captions in the United States). The study found that 7.5 million people in the UK (18 percent of the population) used closed captions. Of that 7.5 million, only 1.5 million were deaf or hard of hearing. This suggests that 80 percent of television viewers used closed captions for reasons other than deafness, and that captions benefit many more than just those who require them for accessibility. When participants were asked why they use subtitles, respondents explained that they are generally regarded as very effective in enhancing understanding of the television programs. The study also reported the benefits of the subtitles to people with attention deficits, autism, and learning disabilities as well as helping with comprehension of dialogue that is spoken very quickly; dialogue with accents and dialects, mumbling, or background noise; and when video includes full names, brand names, or technical terminology. Captions also allow viewers to watch videos in sound-sensitive environments, such as offices and libraries.[13]

In another survey on movie captioning, 5,000 responses were received, more than nine out of ten from the United States and 6.4 percent from Canada. Of these respondents, 38.5 percent were deaf, 15.7 percent were hard of hearing, and 45.8 percent were hearing. Three fourths of all respondents said they prefer open captions (those captions visible on the screen to everyone), and 91.5 percent said that open captions do not bother them.

Some of the respondents believe that as open captions become more common, they will become a norm and less distracting. Some suggested the captions would be less distracting if placed below the moving image.[14]

In describing the attitude problems, Kouznetsova, an accessibility consultant who helps businesses make products, services and events more accessible, wrote that "many theaters make wrong assumptions that open captions on a screen would turn away many hearing viewers." These assumptions are made, she explained, "simply based on a few hearing people complaining about how open captions are 'distracting' to them."[15]

> Theaters are more than happy to fix issues of a missing sound, but many of them don't care much about the issue of missing captions. The incredible example is a theater in UK [that] removed captions from the only accessible movie screening to deaf people to satisfy whims of some hearing viewers who didn't bother to check that the movie was open captioned and could have picked another screening that was not captioned![16]

The Services for Students with Disabilities at Virginia Tech has questioned the validity of arguments that captions appearing on a movie screen are distracting to viewers. "Most persons adapt very easily to the captions and find that they use the captions to supplement the audio when they miss some dialogue due to poor sound quality/acoustics or due to environmental noises. Additionally, research shows that, contrary to the supposition that captions are distracting and interfere with learning, there is no significant difference in test scores between hearing groups who viewed captioned and those who viewed non-captioned media."[17]

## A GLIMMER OF HOPE

In 2007 and 2010, Regal Entertainment Group, on behalf of the National Association of Theatre Owners, conducted symposia in Washington, DC. Attending were representatives from major organizations serving deaf citizens that tested new technological devices and provided feedback to their developers, such as Doremi and Sony.

Progress has remained mixed, however, and the media has carried stories of the frustrations as well as the breakthroughs. As recently as 2016, Katherine Bouton wrote about how movie watching can be tough for those who are deaf. She was on vacation in Florida in a town with many older people. The independent cinema had captioning equipment for deaf and hard of hearing people and she described in 2016 a "bulky device" that had a small screen to display the text. She explained: "It was a chilly day, and she decided to go see a movie, *The Big Short*. "Unfortunately, the captioning equipment was not available. Apparently a part was missing. The manager instead offered me some headphones to use, which I decided to try. About 15 minutes into the movie, the man behind me tapped me on the shoulder and said furiously, 'Take those things off!'"

"I guess they were too loud," Bouton wrote, "but how am I, a deaf person, supposed to know that?" She left her seat in the theater, returned the headphones, and started to walk out the door. But then she turned around, went back into the theater, tapped the man on the shoulder, and said very loudly, "'Those are headphones to help me hear. I'm deaf.' I hope he was humiliated, because I certainly was."

"The rudeness, in retrospect, was understandable," she wrote. "It's the technology that infuriates me. Why do I have to put on heavy, ill-fitting headphones that not only don't work for me, but can distract people around me? It's like installing a wheelchair ramp that not only doesn't get you where you need to go, but blocks the stairs for everyone else."[18]

Some hope arose in 2010 when the CaptiView system was introduced as another closed captioning effort. The system also consisted of a small captioning display unit with a flexible support arm that fit into the cup holder of the theater seat.[19] "As a deaf person," wrote Gary Kerridge in 2013, "I can speak for many deaf colleagues. . . . [that this system] to deliver closed captions to the deaf is, to put it mildly, a dud!"[20] The screen displaying the captions was meant to be positioned in front of the deaf viewer's eyes, but the deaf viewer must constantly refocus from the big screen to the device. This required intense concentration and multitasking. Deaf viewers have complained that this is exhausting and can result in severe headaches and eyestrain.[21]

In addition, the CaptiView device was inaccessible for many people. Movies were often shown at off peak times and choices were limited. "What is more, there are only six devices per screen meaning people who are deaf cannot go to the movies as a group of more than six people. I am just one person, hoping my experiences will illustrate how frustrating it is to see movies advertised as being closed captioned, but having the technology fail when I am sitting in the theatre."[22]

Jackie Andre continued, "If you truly believe what is written on your website—that you welcome people with disabilities, and that you are proud to offer captioning devices—then you need to step up and make these more than empty PR statements."[23]

Currently, many movie theaters offer closed captioning glasses for patrons who are deaf or hard of hearing. These devices were found to be unreliable due to battery failure and lack of proper connectivity.[24] They are often not cleaned properly, posing a possible health risk for those with compromised immune systems. The glasses are uncomfortable and difficult to use, especially for those who use everyday glasses, hearing aids, or cochlear implants.[25]

"Sony's Entertainment Access Glasses seem like something straight out of a science-fiction flick," wrote Claire Boston of the Associated Press in 2014, "the oversized glasses superimpose holographic captions straight onto a movie screen." She described the captions as bright and clear to anyone wearing the glasses and do not distract others in the theater. "They aren't as stylish as other wearable tech such as Google Glass, but Entertainment Access Glasses are the latest advance in closed captioning devices for moviegoers who are deaf or hard-of-hearing."[26]

Boston described the 3-ounce glasses as "large enough to fit over standard eyewear, wired to a small transmitter box the user wears around the neck that appear to beam captions onto the screen," importantly, "right in a viewer's line of sight."[27] The location of the captions can be adjusted by shifting the movable lenses up or down. But many deaf moviegoers found problems with them as well. "I hated them," James "JJ" Jones recalled. "I had to hold my head still throughout the film to see the captions, and if I moved my head a little, I would lose track of the captions and miss some of the dialogue."[28]

According to Erik Nordlof, the problems with these devices are plentiful: battery failure, and lack of connectivity or sporadic connectivity, which leads to dropped dialogue. He says that any theater with those kinds of audio issues would likely go out of business. "The challenge is that deaf moviegoers are not numerous enough to vote with their wallets in this way, so we feel stuck with these devices," he says. "There is a running joke in the deaf community to have a drawer full of ticket vouchers, reflecting the number of times we've tried to go see a movie, experienced an issue with the device, and received a voucher for our troubles."[29]

The solution, Erik Nordlof says, is open captioning, which displays the captions on the screen for everyone, including dialogue, sound effects, and music. In 2014, Matthew S. Moore, president of MSM Productions in Rochester, New York, and the advocacy group We the Deaf People, told a reporter through email, "I prefer open captioned screenings. . . . Electronic devices do not make me feel equal."[30] And Dean DeRusso also felt that the off-screen assistive devices brought deaf and hard of hearing moviegoers "further and further from an integrated theater for all people since the silent movies ended in the 1920s."[31]

## MANY OTHER BATTLES FOR ACCESS

In November 2018, the NAD and SouthSide Works Cinema in Pittsburgh, Pennsylvania, announced that open captioning would be available upon request for deaf and hard of hearing people. This was made possible through cooperative efforts by the parties after a lawsuit was filed by Kenneth DeHaan, a deaf moviegoer who filed an ADA lawsuit alleging that denial of captioning access at the Pittsburgh movie theater was discrimination based on disability. The parties initially agreed on a trial period of open captioning at the theater, and SouthSide Works agreed to make its open captioning availability on a permanent basis. DeHaan praised SouthSide Works for making the changes, saying, "In light of equality in this day and age of available accessibility, I am pleased with the progressive outcome ensuring that every paying customer will be able to enjoy the magic of the cinemas together." Howard A. Rosenblum, CEO of the NAD, which represented DeHaan along with Stein & Vargas, LLP, said, "The NAD applauds SouthSide Works for being willing to show open captioning at any time upon request, and setting an example for the rest of the movie theater industry."[32]

These are just a few examples of the NAD's advocacy. The NAD was also a leader in the regulatory negotiation process in the Department of Transportation, taking steps to mandate increased access to flight entertainment on aircraft so that the movie viewing experience could be inclusive for all passengers.

Hawaii passed a law effective in 2016 requiring the larger theater chains in that state to offer some open captioned showings. The theaters cooperated, he explained, and the open captioned showings were at different times, not all at off-hours. And even though the law expired at the end of 2017, the theaters continued to offer some open captioned showings. Waldo summarized that other theaters in other parts of the country are doing the same:

> So even though the legal battle may have been lost, the public-relations and customer-service approaches appear workable, and in fact, appear to be gaining ground. . . . There are ways captioning can be made to work for everyone. A typical movie is shown 31 times per week – four showings each weekday, five each weekend day. (Friday is a weekend day for movies.) People with hearing loss represent about 8% of the United States population, and we vastly prefer open captioning. Would it be too much to ask theaters for two open-captioned showings at each auditorium each week? And would it be too much to ask that those showings be staggered, so that there is essentially always at least one open-captioned movie at the multiplex, and at least one opportunity to see each movie on a weekend?. . . . Having been involved in movie theater captioning

> litigation for the past ten years, I'm happy to see a project of this magnitude finally coming to a reasonably successful conclusion. We'll be able to have captions for essentially every movie, although there are a few small independent films that may still be released without captions, and there's the persistent, vexing and unexplained problem of uncaptioned previews. But with the adoption and implementation of the DOJ regulations, the theater doors are largely open to us once again.[33]

## THE NAD'S EFFORTS TO IMPROVE ACCESS ONLINE

In March 2011, Donald Cullen sued Netflix, citing their lack of closed captioning for streaming video as a violation of the ADA. A district court judge threw out his lawsuit in 2013. The US Court of Appeals for the 9th Circuit then upheld that decision. The decision was "unpublished," meaning it isn't intended to be used as precedent in other cases. Joe Mullin summarized that "it certainly doesn't bode well for any plaintiff thinking about filing a similar case in the 9th Circuit, which covers most of the Western US."[34]

However, three months later, the NAD filed an ADA lawsuit against Netflix in Massachusetts over the same issue—the lack of captioning on Netflix programming. The captions were again found to be awkward, some containing actual mistakes. "While I can't speak for viewers who are deaf or hard of hearing," Jon Christian wrote, "I've been watching Netflix with captions enabled for a few months, and every error pulls me out of the experience like a typo in a hardcover novel."[35]

This time the judge found that Netflix was a "place of public accommodation" and would have to face the lawsuit against the NAD. After the company lost the initial motion, Netflix settled the case with the NAD, agreeing to pay $750,000 in legal fees and caption all of its videos by the year 2014.

The subsequent response by Netflix helped set the trend for others involved with video streaming services. In 2014, Netflix stated, "We need to change the way we think about subtitles and closed captions. They are no longer secondary assets in a world where content knows no physical borders." Netflix also began audio description[36] for blind and low vision individuals. By 2015 American Printing House's *Access News* reported eighty-seven programs with audio description offered by Netflix.

Years later, the NAD sued Harvard and MIT over their free online course offerings, saying the lack of captions constituted an ADA violation. A settlement was reached in this case.[37] On November 27, 2019, the NAD announced a landmark settlement with Harvard University. New guidelines were instituted to make the university's website and online resources accessible for deaf or hard of hearing users. This included the guideline that captions for all online resources be made available for schoolwide events that are "live-streamed, content from department sponsored student organizations and any new audio or video hosted by third-party platforms such as YouTube, Vimeo and SoundCloud."[38] The lawsuit was prompted by the recognition that although Harvard's online resources were said to be available to "learners throughout the world," many of its videos and audio recordings were not offered with accurate captions.

In 2019, the NAD's CEO Howard Rosenblum posted advice to the Deaf consumer. He first explained how federal law requires that movie theaters provide captioning through closed captions. Many years ago, the NAD had asked DOJ to add open caption requirements in addition to the closed caption requirements. The DOJ refused this request and decided that closed captioning was good enough for access. "But many of you are frustrated with [closed captions]," Rosenblum signed, "because the [open captions] technology isn't perfect: sometimes the [closed captioning] devices miss some lines, the glasses aren't comfortable, the captions were wrong for the movie used, the [closed captioning] device gets disconnected, or the battery dies. All these issues are good reasons why many of us want [open captions] at the movie theaters." He continued:

> For that reason, the NAD recognizes and commends Hawaii for being the first state to pass a law requiring opened captions (OC) showings of movies at movie theater in 2015. However, later in 2017, legislators in Hawaii amended the bill to allow movie theaters to provide either [open captions] or [closed captions] with movies. This meant [open captions] usage decreased. The deaf community in Hawaii shared their concerns with the legislators to change the bill back to requiring movie theaters to show movies with [open captions]. This did not mean no [closed captions], only that people can have both [open captions] and [closed captions]. Hawaii isn't alone in this effort, there is also a

push in the DC legislature to require [open captioned] movie showings at movie theaters. A few other states are doing the same.[39]

Rosenblum encouraged people to advocate for their state legislatures to pass laws for "what we want—OC movies on a weekly basis in movie theaters! Fight for more options not just weekday morning movie times."

He explained that people who live in states that do not have open caption requirements have a right to file a complaint about a closed caption experience if they are not able to understand the movie. The NAD offers an online form. Once the form is submitted to the NAD, it will be passed on to the movie theater chain. "We can only achieve equality if all of you advocate for your rights."

## TDI'S DECADES OF CAPTIONING ACTIVISM

Telecommunications for the Deaf and Hard of Hearing, Inc. (TDI), has this as its mission statement: "To promote equal access in telecommunications and media for people who are deaf, hard of hearing, late deafened, or deaf blind."[40] TDI has been a key player in this history of captioning, particularly with regard to implementing captioning technology in the Deaf community and advocating access through FCC consumer education and involvement, technical assistance and consulting, networking and collaboration, uniformity of standards, and national policy development.[41]

In 1972, TDI began its advocacy of captioning by attending the first National Conference on Television for the Hearing Impaired in Knoxville, Tennessee. Through the 1970s, under H. Latham Breunig, TDI joined the Council of Organizations Serving the Deaf, a coalition that met to discuss pending federal legislation that could have an impact on deaf Americans. As the captioning industry subsequently emerged, TDI's agents, who first helped deaf consumers install teletypewriters for telephone communication, expanded their role by installing TeleCaption decoders in homes and businesses. This included training on how to use the new technology.

In 1982, TDI began selling NCI TeleCaption decoders. Throughout the 1980s, topics related to captioning were offered in the Technology

Forum of TDI's annual conventions. These meetings often included activities such as presentations by FCC Chairman Reed Hundt, and workshops and fora on captioning technologies such as Clear Captions (Mike Strecker) and CaptionCall (Cameron Tingey), general TV captioning issues (Lise Hamlin), and closed captioning rule making (Rosaline Crawford). Through the 1980s and 1990s, TDI executive directors Barry M. Strassler (1979–1983), Thomas M. Mentkowski (1984–1986), and Alfred Sonnenstrahl (1987–1996) oversaw training of agents to install captioning decoders, continued decoder sales, and advocacy for accessible telecommunications and increased captioning, and advising consumers on the purchase of televisions with the built-in captioning chip. As Chairman Tom Wheeler of the FCC later summarized at the 2015 TDI conference, during Sonnenstrahl's fifteen years with TDI, "Sonny" "helped spearhead many of the giant advances in telecommunications access—for example, access to 911, the ADA's requirement for relay services, and closed captioning. He also played a role in the FCC creating its Disability Rights Office. "Telecommunications equality would not be what it is today without [his] extraordinary work during those early years and we thank you for that."[42]

Under Claude L. Stout (1997–2020), TDI submitted numerous filings to the FCC in such areas as digital television captioning, increased accessibility of video programming to viewers with hearing and vision disabilities during local emergencies, and internet captioning. When evidence surfaced that the FCC was not following its own protocols for reviewing requests in the Angler's Exemption Order to grant captioning waivers to provide captioning on TV, TDI and other organizations filed an application for review, resulting in further modifications of policy.[43]

On October 17, 2007, Stout presented testimony to the House Subcommittee on Technology and the Internet, chaired by Representative Markey, about the difficulties in accessing captioning on new digital television sets. Then, in a historic first for the TDI executive director, he testified on behalf of the Coalition of Organizations for Accessible Technology about digital television transition issues with captioning. The FCC then began a campaign to promote a digital-to-analog converter box coupon program for viewers still using analog television sets.[44]

TDI became increasingly involved in captioning advocacy during the new millennium. Among the numerous activities described in

TDI's World Anniversary Commemorative Edition, *A Path toward an Accessible World: 1968–2018*, was the series of meetings between TDI and other consumer groups and the US Department of Transportation, which led to the release of the Airline Carrier Accessibility Act's first major regulatory overhaul. This work resulted in positive changes for travelers who are deaf and hard of hearing, including captioning for in-flight entertainment.

TDI also filed comments with the US DOJ in several advanced notices of proposed rulemaking relating to updating the ADA regulations on movie captioning and video description; the ADA regulations on accessibility of websites operated by state and local government offices and public accommodations; the ADA regulations on equipment and furniture, specifically those that provide electronic information technology, such as access to information kiosks and other communication devices like restaurant drive-thrus, ATMs, building intercom systems, and other audiocentric communication systems; and the ADA regulations on access to the Next Generation 9-1-1 access to emergency services operated by state and local government entities.[45]

Between 2012 and 2016, comments or oppositions on behalf of TDI with the FCC concerning more than 130 petitions, leading to the dismissal or denial of more than 60 petitions that TDI opposed. In 2016 TDI received a $500,000 two-year grant from Google for a captioning correction project.

These are only a few of the many accomplishments of TDI related to captioning. The organization's ongoing collaboration with other consumer groups serving people who are deaf or hard of hearing has been "a force multiplier that leverages its technological expertise and its vision that anything is possible when new and emerging technologies are equipped from the very beginning with accessibility features such as captioning, text, and video along with traditional audio information."[46]

## ONLINE CAPTIONING ADVOCACY DISCUSSION GROUPS

In addition to the captioning advocacy of such organizations as the NAD, TDI, Described and Captioned Media Program (DCMP), and the individual and smaller group efforts on the state and local levels, there have been some outstanding online discussions of captioning.

One of the earliest of these groups was on Yahoo. Begun in the late 1990s by Jamie Berke, the group's coordination was continued by David H. Pierce and grew to 1,402 members. It remained active until December 15, 2020, when all Yahoo captioning discussion groups were ended. Catalin Cimipanu explains that over time, the "once-massive userbase" was lost to newer services like Reddit, Google Groups, and Facebook Groups.[47]

Jim House, a former staffer at TDI, who was a regular participant in the Yahoo group discussions, has been running a Facebook group called Captioning Crusaders.[48] The advocacy efforts shared among this group's members are wide ranging and include topics such as these: providing free automatic speech recognition captions for deaf people on video conferencing platforms through Zoom, Google, Microsoft, and other video conferencing providers; the DCMP captioning philosophy and its *Captioning Key;* the fight for closed captions on in-flight entertainment; mobile phone accessibility; and highlighting the accomplishments of Julius Barnathan, one of the original behind-the-scenes revolutionaries who headed broadcast operations and engineering for ABC.

The Collaborative for Communication Access Via Captioning (CCAC) has an online Captioning Advocacy forum for CCAC members that is part of the CCAC nonprofit organization.[49] Lauren Storck started the organization and discussion group online in 2009 and it rapidly grew to hundreds of members. In comparison to social media groups, CCAC awards grants for inclusion of quality captioning to nonprofits that have never included captioning access previously, and it conducts a variety of captioning inclusion campaigns, including a large national cinema captioning project.[50] The CCAC website summarizes a wealth of captioning information and advocacy activity. All staff are volunteers. Within CCAC's mission, "Inclusion of Quality Captioning Universally," citizen advocacy is the goal, whether it's a small local need or wider national and international issues, to ask for and achieve captioning needed in so many places. In the online forum, people share captioning news, information, and advocacy questions and solutions.

# 16

## Spotlight on Technology for Educational Films

*Captioning and description are literacy tools to benefit millions of others, and this cause is championed by the DCMP through its Read Captions Across America and Listening Is Learning campaigns.*

—Jason Stark, DCMP director

In 2011, the NAD won another five-year cooperative agreement to continue management of Described and Captioned Media Program (DCMP), with an annual budget of $1.5 million. Jason Stark became the new project director, and his father, Bill Stark, moved to a half-time DCMP position and semiretirement. Bill had been the NAD's DCMP project director for more than twenty years. Before retiring in 2017 he also served as communications and accountability administrator for DCMP, focusing on client communications and working with DCMP staff on a broad range of client recruitment and promotional activities.

Prior to serving as project director, Jason's job duties focused on instructional technology and distribution management. He began his career in information technology (IT) as a high school student, conducting computer-use workshops for special education personnel who affectionately dubbed him the "Wizard."[1] Since 1995, when Jason moved from selection/captioning to distribution of captioned films, he had been responsible for overseeing technology for both projects. He immersed himself in website accessibility issues, consulting with programs across the country in a mutual quest for equal access, and initiating collaborations with technology giants such as Google.

Jason was at the forefront of online educational applications, and by 2018 the DCMP website had more than 20 million hits. As a result of his efforts, the project under the NAD now provides accessibility information available through the website's Learning Center with more than 300 original articles on aspects of captioning, description, education, accessibility, deafness, and blindness.

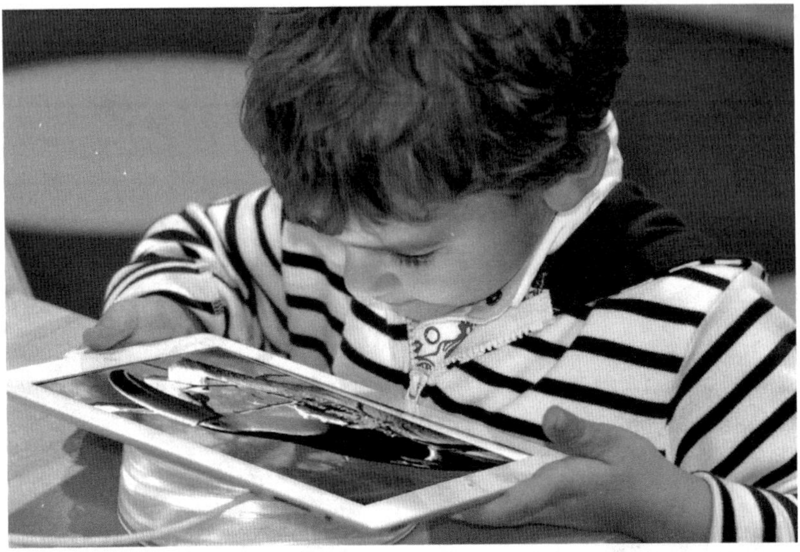

DCMP's many resources for captioned and described media provide opportunities to learn on a variety of devices in the home and school environments, including desktop computers and a variety of portable devices such as tablets and laptops.

## EXPANDING THE USE OF SOCIAL MEDIA AND STREAMING SERVICES

DCMP also launched a YouTube channel in 2008, with the goal of utilizing this outlet in promoting the program, advocating for accessibility, and providing accessibility information. By the end of 2019, DCMP had reached 22.4 million total views on its YouTube channel.

DCMP staff also teamed with Google to provide an evaluation process to assess captioning quality for YouTube vendors. Some of the vendors were resistant, but Google was very supportive of the DCMP's efforts to bring quality control. DCMP was not involved with the machine-generated automatic captions, which received criticism for not being in full compliance with the ADA.[2] However, viewers like Rikki Poynter, who created "No More Craptions," have worked hard to address this problem, which continues to exist today.[3] In an article titled "The Problem with YouTube's Terrible Closed Captions," it was summarized that "craptions" have improved, with fewer inaccuracies. However, stringent requirements are not enforced, and businesses failing to provide adequate captions may face a lawsuit. Individual social media creators, however, are unlikely to be found in violation of the

ADA. James Rath, a YouTuber and filmmaker, explains that captions can expand the reach and performance of a video. Search engines may pull keywords from the transcript.[4]

In 2009, DCMP launched its Facebook page, making it easier for clients and others to track program activities and services. Deaf, hard of hearing, blind, visually impaired, and deafblind students have enjoyed the many resources described and captioned, including many videos. Captions have helped deaf and hearing people overcome reading disabilities. Some would not have known the benefit but for the challenge presented by deafness to family members or friends.[5] Twitter is also used to promote the program's services. These social media platforms have greatly enhanced the program as a central resource for educators of deaf students throughout the country.

After Jason became project director, he also led DCMP in launching a channel on Roku, a streaming set-top box with thousands of channels including Netflix, HBO, Showtime, ESPN, and History Channel. Roku, which uses existing Wi-Fi with a HDMI connection to a television or projector, features fully accessible educational media designed for use in the classroom or at home with a remote control.[6] DCMP offered the only channel in the nation with fully accessible

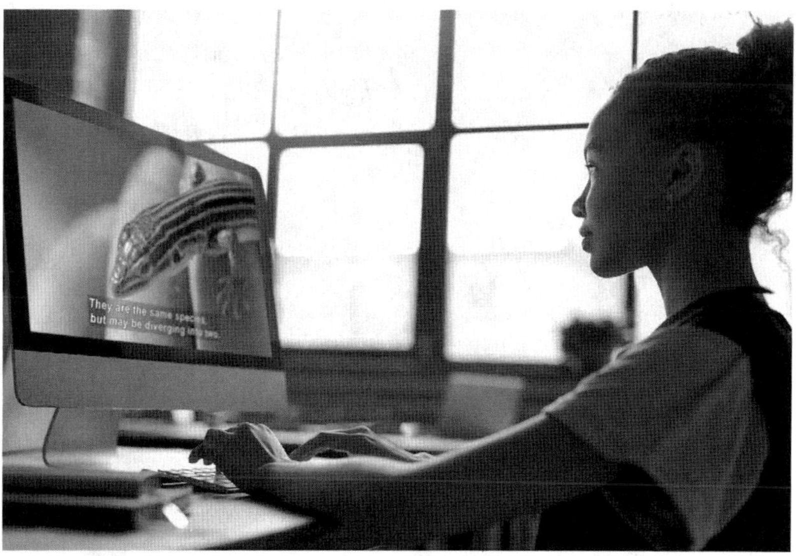


A girl uses a desktop computer with captions.

educational material. All video content can be embedded into a member's LMS (learning management system) or CMS (content management system, such as WordPress), or virtually any other HTML-based web presentation. In 2015, an app was developed for Apple's iPhone, iPod Touch, and iPad users, and it is offered as a free download to DCMP members.

DCMP's many additions required quality control. From 2010 forward, NCI and many independent vendors captioned for TV and others, but in order to work for DCMP, a test had to be passed to prove that the *Captioning Key* guidelines were followed for quality.

**White House Anniversary of the Americans with Disabilities Act**

In August 2012, Karen Peltz Strauss, then deputy bureau chief, Consumer and Governmental Affairs Bureau, was honored to represent the FCC at the White House's observation of the ADA's twenty-second anniversary. Joining a panel of four other administration officials, she spoke on the many ways in which the Commission had successfully implemented the Twenty-First Century Communications and Video Accessibility Act (CVAA), signed by President Obama in October 2010. The CVAA requires access by people with disabilities to emerging internet-based and digital communications and video programming technologies.

Strauss highlighted four areas of accomplishment by the FCC, including the FCC's adoption of the rules in April 2011 and the launching of the National Deaf-Blind Equipment Distribution Program on July 1, 2012. The rules required video description to be provided on approximately four hours of prime time or children's television programming each week by local TV station affiliates of ABC, CBS, Fox, and NBC in the top twenty-five TV markets as well as on the top five nonbroadcast networks—that is, Disney, Nickelodeon, TBS, TNT, and USA. Then, in October 2011 the rules were adopted requiring advanced communications services (such as email, instant messaging, and VoIP communications services) and the products used with those services to be accessible to people with disabilities. In January 2012 the rules were adopted that require closed captioned television programs to retain those captions when they are re-shown via internet protocol.

—Karen Peltz Strauss, "White House Anniversary of the Americans with Disabilities Act," FCC.gov, August 2, 2012, https://www.fcc.gov/news-events/blog/2012/08/02/white-house-anniversary-americans-disabilities-act

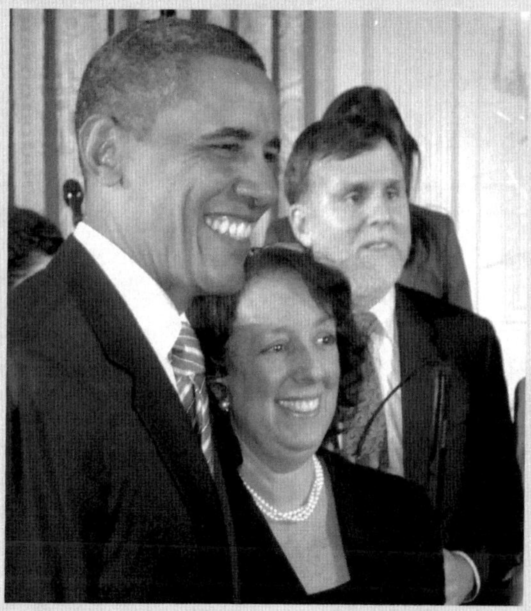

Karen Peltz Strauss and President Barack Obama.
Image courtesy of Andrew Kirkpatrick.

In 2012, JoAnn McCann became DCMP's project officer in the US Department of Education, replacing Ernie Hairston.[7] She has been a strong advocate for DCMP and has also been in charge of the TV access grants since their beginning, in 1990, after the decoder chip law was passed. At that time, the government began funding captioning of TV in order to increase the hours of programming. McCann managed the access grants, the content of which fuels the DCMP portal. The progress would not have been made without McCann's support, and she was rightly recognized by the Department of Education for her role in it. The Golden Apple award is presented to an employee who demonstrates outstanding service of local, state, national, or international significance. In 2015, McCann, now in the Office of Special Education and Rehabilitative Services, was honored for brokering a deal with television network executives to give DCMP the rights to add television programming to its accessible portal for teachers and students who are deaf, blind, hard of hearing, visually impaired, or deafblind. Through McCann's vision and leadership, seventy-three episodes of nineteen different children's television programs were made accessible to children and teachers with disabilities. Upon McCann's retirement in 2018, Glinda Foster Hill became the project officer for DCMP.

DCMP continued to experiment with new technological approaches to provide educational film access through captioned and described media. In 2014, the FCC took action to improve captioning quality on television by putting into place minimum requirements for the accuracy, synchronicity, completeness, and placement of captions displayed on all television programming covered under the FCC's rules.[8] DCMP's *Captioning Key* is referenced several times as an exemplary set of captioning standards in the FCC's Report and Order that adopted these guidelines.[9] The *Key* had originally been developed using manuals from CFD as well as from NCI. Some vendors were new, and DCMP trained their staff members to caption.

With the aim of increasing the accessibility of captioned media for children who were deaf, DCMP also partnered an initiative with the National Education Association (NEA) called "Read Captions Across America." The goal, based on educational research findings, was to provide NEA members, parents, caregivers, and children the resources and activities they need to keep reading throughout the year. DCMP's website provides captioning information, a database for locating educational captioned media, and other forms of support.[10]

DCMP also partnered with CaptionMax in that company's two "Emerging Technologies" federal grants. Products of the grant activities were distributed to DCMP members. In particular, CaptionMax provided some exciting new features on DVDs. One innovative feature explored was interactivity in captions, which allowed words and phrases to be "clickable." The new interactive transcript tool was used in conjunction with the core media player. While viewing DCMP's streamed media, users were able to search the captions (transcript) to quickly locate a specific concept, and then, with one click, could jump to that point in the video to view the segment. The new tool also allowed members to scroll through the transcript.

Viewers also had the choice of using either verbatim or edited captions on the educational films. For viewers who were blind or visually impaired, autoplay spoken menus and voiced navigation instructions were provided. As an experienced and trusted provider of captioning and video description of educational media, CaptionMax is committed to making high-quality educational television fully accessible to blind, low vision, deaf, and hard of hearing children, parents, and educators. Since 1993, thousands of hours of programming have been captioned and video described, including preschool and early elementary children's programming that is broadcast nationally.[11]

In 2016, the NAD again was awarded a five-year grant with annual budgets increased to $2M a year under Jason Stark. DCMP's YouTube channel amassed more than 11 million lifetime views at this time. The NAD serves 124,000 members with a collection of 5,500 accessible titles (now including 886 captioned and/or described titles in Spanish). More than 95 percent of all program user transactions are of streamed media, with 5 percent being DVDs delivered by mail. From a limitation in 2002 of seventy simultaneous users, there was in 2016 virtually no limit to that number.[12] As much as possible, the DCMP staff has attempted to follow universal design as its guiding principle. That is, the guiding principle should be that the design and composition of an environment (the building, product, or service in that environment) "can be accessed, understood, and used to the greatest extent possible by all people regardless of their age, size, ability or disability."[13]

## MORE ENHANCEMENTS IN DCMP

For the first time in the history of the program, deaf (and blind) students were allowed direct student access to its captioned and described media content. Student accounts were created to support teachers and parents guiding and mentoring students in the learning process.

Ever since PL 94-142 was passed in the mid-1970s, there has been a concern about ensuring that deaf and hard of hearing students become knowledgeable about captioned films. Scattered all over the country in different school programs, these students and their teachers may not be well-informed about the rich resources in the DCMP. Significant discussions about this were a major item for the program throughout the years, and they continue. Then, in 2017, Jason Stark received approval to hire a full-time employee dedicated to marketing. The internet has been the best marketing tool and a boon to helping address this issue. Deaf children and their families often have never heard of DCMP, but a quick Google search often connects them with the necessary resources. Sometimes, a specific video title may be found by parents or teachers on the DCMP website, and after navigating a bit, they learn about the huge library of captioned (and described) videos they can borrow once they register.

As teachers used media more often as an integral part of their curriculum, it became more common for captioned media to be included in a deaf student's education plan at the middle and high school levels. Less attention was given to captioned media at the elementary level, in part because videos and film clips were not used as often. DCMP increasingly became an integral component of educational planning.

Some vestiges of the original program were eliminated at the time Spanish captioning was added, including the national library of VHS videos. In 2006, the project had started producing DVD copies of materials and distributing them through the full library network. DVD distribution, however, was a challenge due to the frailty of the disc—they came back scratched and broken, and the staff at the depositories were not equipped to handle the quality control. Primarily due to the budget cut, but also due to DVD and streaming technology taking off, a decision was made to pull all off-site distribution from the libraries. Educational media were henceforth shared through streaming and DVD copies made available from DCMP's main office.

Quality television shows were made accessible through recipients of five Department of Education–funded television access grants. During this time of transition, DCMP began to partner with these grantees and their content partners to make their accessible content available to DCMP's members, distributing it through an accessible television portal on DCMP's website. A range of other new technologies served to modernize the captioned and described educational film support for school programs.

DCMP also partnered with the American Council of the Blind in sponsoring the Young Described Film Critic Contest. Through these efforts, which began early in the 2006–2011 budget cycle, many thousands of children who are blind or have low vision boosted their literacy skills by listening to described programs. In 2016, this program was rebranded Benefits of Audio Description in Education (BADIE) and continued to work with DCMP to encourage young people challenged by blindness to read more through audio description. They developed the BADIE contest for students to compete for prizes by summarizing their experiences with media using audio description. An audio described film is chosen from the more than 6,000 titles available through DCMP, or a student can borrow an audio described video or film from a Talking Books Library or a local public library. DCMP also launched "Listening Is Learning" as a one-of-a-kind campaign focusing on the benefits of using description both with its intended audience (blind people or those with low vision) and the larger population, specifically in the field of education. The American Council of the Blind and the International Listening Association were recruited as partners. This activity was part of an effort by DCMP to encourage producers to voluntarily add description to their media.[14] Jason Stark emphasized the benefits of described and captioned media: "[C]aptioning and description are literacy tools to benefit millions of others, and this cause is championed by the DCMP through its Read Captions Across America and Listening Is Learning campaigns."[15] As Linda Cannon, family support resource specialist at the Kentucky School for the Deaf, wrote, "The additional programs like 'Read Captions Across America' helped our students participate in a national event and helped spread the word how captioning of educational materials is necessary for our students to achieve to their maximum potential."[16]

## THE BATTLE WITH MEDIA PRODUCERS

DCMP's success did not come without a struggle. Whether with entertainment or educational films, the difficulty in obtaining improved accessibility through captioning and description has been primarily an issue of ignorance and expense. The cost issue has abated significantly over the years. Despite the strong support from consumers and collaborating agencies, DCMP has faced a never-ending struggle to level the playing field with accessible educational media in particular. In 2007, the need for enhanced captioned and described film resources for K–12 programs was made clear in a survey of the top educational media producers. It was found that out of thousands of products available, only about 25 percent were captioned and less than 5 percent were described.[17] In another online survey to measure the awareness of the described media resources and to study trends concerning overall video usage, DCMP solicited input from 222 teachers of students around the United States. Eighty-six percent of the teachers were familiar with the concept of description, and 45 percent used video in their classrooms. Of the teachers who used video, 71 percent reported that very few of the videos they used were available with description.

Thus, it has remained critical to disseminate information about program resources. In collaboration with Jacksonville State University in Alabama, DCMP produced a training video titled *Equal Access in the Classroom*, which was designed for teachers. The video defined description and captioning and provided examples of them, offered teacher testimony supporting the need for these accessibility options in educational media, and overviewed DCMP's services. Additional productions with Jacksonville State University included a series of sign language vocabulary videos in 2010 and a series of videos to teach others how to add captions to already created media in 2011.

Both Bill and Jason Stark played critical roles in the activism needed to increase the availability of captioned and described materials for use with students in American classrooms. The project also has a long history of encouraging consumers to be activists. Back in 1996, Bill Stark and his staff attended the Association for Educational Communications and Technology conference and subsequently assigned several consumer advisors to advocate for access at exhibits and other networking events involving thousands of educators. But, a decade later, in 2006, DCMP still faced resistance to make media accessible. In the late fall of 2006, DCMP organized a large contingent of the NAD staff, DCMP board members, and DCMP staff to walk from the NAD headquarters to Discovery Communications, a stone's throw away, in Silver Spring, Maryland. The plan was to meet with Discovery's key sales and production staff members over lunch, followed by a meeting in a board room, with selected staff members from both Discovery and DCMP. Bill Stark felt he had taken on Goliath—the organization had greatly resisted captioning anything (except what they were required to do so on network TV) and/or allowing DCMP to have much of anything in terms of content. A group of deaf, blind, and deafblind consumers first mixed with Discovery's staff in their cafeteria. As Stark's group gathered for a meeting with their key staff, set up in advance for the purpose of Stark and several of the consumers to make a short presentation about access, the Discovery staff started, in effect, to cancel the meeting. Stark recalled, "Being treated with such disrespect, I stood up anyway in the room and delivered part of the speech I had prepared, as did a couple of consumers. The Discovery staff were "arrogant and rude. . . . and, of course, all of us were disappointed. We were not intimidated, and I dare say it brought us closer together as a group." Stark reflected, "Discovery had empowered us, increased our resolve."[18]

Although the experience with Discovery was a disappointment, activism at the gatherings of the National Media Market were more successful. For many years, media company representatives came to the market to display their catalogs and provide screening opportunities for the media they were selling. Media buyers came to preview, negotiate, and buy content for their schools/universities. Advisors would spend a couple of days previewing media items marketed by the companies and then would visit the suites of the companies to give feedback on the productions and make the pitch for the importance of accessibility.

DCMP's consumer board members networked with producers and distributors, evaluated their media for captioning and description, and advocated with the attendees for access and for them to allow program availability to their content.

Bill Stark recalled how, back in 1993, the head of the market pulled him aside and basically told him that his staff shouldn't be bothering the company representatives. He said that the companies were there to sell their product and didn't have time to discuss accessibility. "It was clear that we weren't exactly welcomed there," Stark reminisced. They nevertheless continued to push the importance of quality accessibility.

The DCMP staff had been attending the National Media Market off and on for many years. Bill Stark fought a long battle with those who would not allow DCMP to include any of their content in their collection. In 2009 he communicated directly with the market's CEO, Abbe Raven, pointing out the reasons why they should change their attitudes about access, and she personally approved giving DCMP the title *Barack Obama*, an episode from the Emmy award-winning *Biography* series from A&E. This was the first step, but progress remained slow.

By 2013, at their annual convention in Charleston, the National Media Market had done a complete turnaround and literally were lining up to make appointments with Jason Stark and other technical staff to learn more about description for the blind, digital advancements in captioning, and so forth. Then, in 2018, Jason was approached by the current chair of the market who *invited* DCMP to present a professional development session to the group on accessibility. Jason recommended Lily Bond from 3Play Media, which provides accessibility services, and Blake Reid, an attorney from Colorado who focuses on accessibility law. At the market, Jason provided information about captions and descriptions, a simulation of the experience of a blind student when watching an educational video with and without description, a quick overview of what makes up quality captions and descriptions, and finally information about some of the tools that can be used to produce accessibility features. Lily Bond presented about automatic speech recognition and how it fails to meet the expectations of quality, and how 3Play and other vendors offer integrations and technological solutions for the generation of accessibility features on a large scale. Blake Reid presented information about copyright law.[19]

Before one presentation at a meeting of media producers, a representative stood up and commented to Bill Stark that because DCMP staff had been coming to the market for years, playing a dual role of accessibility advocate and media buyer (licensor), the majority of companies that had media appropriate for K–12 classrooms were convinced that captioning was necessary. "That is another issue we've faced and overcome," Bill Stark wrote. "The program used to pay thousands of dollars to license content from media producers and distributors." In 2011, the Department of Education mandated that the program pay nothing for content—just offer the captions and descriptions in exchange for duplication rights. It was not easy to convince the companies, but DCMP pulled it off. Due to the continued presence of DCMP staff at the market and the push for access, the companies were seeing the value in its collaboration with the program.

Over the years, advocacy had taken many different forms—not all are connected to a single activity. For example, Weston Woods, a division of Scholastic, produces animated videos based on award-winning children's books. It has been on the forefront of offering accessible products for many years. DCMP believed it was very important that the broadest range of users be able to utilize its products. Weston Woods was one of the first to incorporate closed captioning on all of its children's video titles, even before it was federally mandated. Both they and DCMP remained current with the newest trends in accessible offerings; it was because of this that described audio tracks were added to the company's DVDs along with the closed captioned tracks.

The *Captioning Key* continues to be an important advocacy tool, not only stressing quality accessibility but also providing brand recognition to DCMP. Many companies now link to it; Microsoft at one time used it for their staff. It has been translated into different languages.

DCMP has also become increasingly involved in training, another form of advocacy, offering eLearning modules on various accessibility topics[20] and QuickClasses that instruct teachers how to use accessible media in the classroom.[21] These classes have been approved by the Registry of Interpreters of the Deaf and the Academy for Certification of Vision Rehabilitation & Education Professionals to issue continuing education credits.

What all of these technological innovations by DCMP have done is to very efficiently fulfill the Department of Education mandate to address the needs of K–12 students. With continued funding there is great promise for the program's impact on teaching and learning.

## Epilogue

## The Enigma of Cultural Awareness

*Film is our literature, so we should tell stories that are apropos of our culture, in that we can learn something about ourselves.*

—David Strathairn

"Oh, so you learned how to read and watched *Parasite*? Congratulations!" exclaimed Bong Joon Ho onstage at the seventy-seventh Golden Globes in June 2020. "Once you've overcome the one-inch-tall barrier of subtitles, you will be introduced to so many more amazing films."[1] The celebrated writer/director of *Parasite* would also win the Academy Award for Best Film in 2020.

While watching the Golden Globes and reading this caption, I looked at my wife and she nodded. Bonnie knew what I was thinking. I had been working on this book for a long time. I signed to her, "We are not alone."

Just like foreign filmmakers, who wish to communicate other cultures through non-English films, deaf and hard of hearing Americans have faced the attitudes of those who believe that open captions are "distracting." Some, as this history reveals, have actually interfered with legislative attempts to bring access to films through open captions to deaf and hard of hearing people in theaters.

American culture is connected to film, and even though deaf Americans have fostered a Deaf culture of their own, the difficulty they have had in gaining access to entertainment films has meant missing out on learning opportunities about mainstream American culture. This theme of "cultural awareness" repeatedly surfaced as I conducted research on the history of captioned films. As I have discussed in this book, learning about culture through both educational and entertainment films was the underlying goal that led the US Office of Education to fund Captioned Films for the Deaf in September 1958. PL 85-905, the very first law authorizing funding for captioned films, set forth several purposes. One stated goal of federally funded captioning was to bring to deaf people "understanding and appreciation of those films which play an important part in the general and ***cultural*** advancement of hearing persons" [emphasis mine]. A second purpose was to "provide, through these films, enriched 'educational and ***cultural*** experiences' through which deaf persons can be brought into better touch with realities of their environment." The third purpose was to provide a wholesome and rewarding experience that deaf persons may share together. In his memoirs, written a year before his death, John A. Gough mentioned that in the 1960 conference to kick off the plans for implementing CFD, "Among the important outcomes of this meeting was to be a quieting of the fears felt by some leading deaf persons that placement of the film program in the Office of Education would tend to obscure the recreational and cultural objectives which they held to be of primary importance."[2]

Gough's concern was prescient: Government funding *was* eventually eliminated for entertainment films due to changing priorities in the education of students with disabilities on the national level.

The more I read about the subsequent struggles of deaf people to access entertainment films in theaters, the more I wondered about the Department of Education's initial goal of developing an "understanding and appreciation of those films which play an important part in the general and ***cultural*** advancement of hearing persons." It seems to me that American society, after all these years, continues to make the sharing of culture quite difficult for deaf and hard of hearing people.

In studying captioning over the past few decades, I noticed that even a film with a strong orientation to the cultures of Deaf and hearing people had met with access problems. In 1986, the film *Children of a Lesser God* dealt in part about the cultures of deaf and hearing people and did much to enhance the understanding of deaf people, their culture, and American Sign Language. It opened the minds of many hearing moviegoers about the experiences of deaf people. As one reviewer wrote, the two main characters, James (hearing) and Sarah (deaf), could not "experience what the other experiences every day and this becomes a type of conflict."[3] The benefits of this film were profound. As a deaf person who personally experienced the effects of this film, I was thrilled. Waitresses in restaurants began to learn to sign in high school classes. Friends were showing increasing curiosity about Deaf culture. "Deaf Studies" as a field was growing rapidly. There was much more respect for ASL as a true language.

Ironically, deaf and hard of hearing people had great difficulty comprehending this film when it was first released. Paramount Pictures was not equipped at that time to handle the undiscovered market of deaf movie fans. There *was* sign language in the film, but the film frequently cut away from the sign language communication. Without captions, the film was largely inaccessible to the deaf audience. Of the 215 theaters showing *Children of a Lesser God*, only ten had captioned versions. And because these captions were screened only once a week, often on a Saturday or a Sunday (never in the evenings), most deaf people who wished to see it were out of luck. Paramount's excuse was that the captions were too expensive.[4]

The character Sarah was played by the Deaf actor Marlee Matlin, who has been a strong advocate of captions over the years. On February 9, 2016, Matlin wrote to DCMP director Jason Stark that "Captioning is of paramount importance to me in that it makes all forms of video media accessible to me but also to 48 million deaf and hard of hearing people in this country." For her 1987 film debut in Paramount Pictures' *Children of a Lesser God*, she received the Academy Award for Best Actress. She has devoutly supported captioning access, the NAD, and DCMP in their efforts to provide and promote equal access in the classroom.

## A NEW CULTURAL PHENOMENON

One cannot read this book without asking whether there are data supporting the claims that captions are "distracting" to hearing people. Can moviegoers from the two distinct cultures not share the medium of motion picture films in a theater house for a few hours? Many hearing people responding to surveys say they are not distracted by open captions. In fact, there are numerous reports on the benefits of captions to all viewers. Access through closed captions on television is now a requirement. Many hearing people turn them on for various reasons discussed in this book.

But at the time of publication, access for deaf and hard of hearing moviegoers is still not equitable to that of hearing people. The films are often not released to premium cable channels, streaming services, or a pay-per-view option until months after they are shown in theaters.

Finding those films through libraries or supermarket kiosks is also difficult.

As this history has revealed, the attitudes go far beyond the issue of "distraction." In the past, educational media companies resisted captioning their products, and filmmakers and film vendors resisted efforts to caption their films. Hollywood resisted efforts to pass laws to require open captions in theaters, and theaters resisted the installation of captioning technologies. The Department of Justice resisted

**Karen Peltz Strauss—A Legend**

In January 2019, Michael Copps, who began as commissioner of the FCC in 2001 and served for eleven years, gave a presentation titled "A Stellar Public Servant." He stated that "A legend is leaving the FCC as the new year begins. Her name is Karen Peltz Strauss. Some of you may not have heard of her, but to the nation's disabilities communities, she is a hero. She achieved this status the old-fashioned way. She earned it. . . . "

Among Strauss's many accomplishments summarized by Copps was how, before joining the FCC, she had been instrumental in formulating critical legislation like Title IV of the ADA (requiring telephone relay services) and Sections 255 and 713 of the Communications Act, which mandated telecommunications access and television captioning. "She never stopped," Copps said. "As part of the effort to enact the 21st Century Communications and Video Accessibility Act, she founded a coalition of some 300 national and regional organizations whose grassroots efforts were essential to passage of this historic legislation."

Strauss also helped create the FCC's Disability Advisory Committee, a body that provides a "valuable forum for the exchange of ideas among consumer stakeholders, industry, and researchers. It's the kind of thing that helps make the government, and the nation, work."

—Michael Copps, "A Stellar Public Servant,"
*Benton Institute for Broadband and Society*,
https://www.benton.org/blog/stellar-public-servant

efforts to mandate open captions in public theaters.[5] Although closed captions were required, the technology has not been satisfactory, and most deaf moviegoers prefer open captions.

Hence, the story of captioning told in this book turns into one of not only outstanding technological breakthroughs, legislative victories, and leadership in the government-sponsored captioning program, but also a demonstration of the need for advocacy on many levels.

This history of captioned films has paralleled in time the activism that was needed to gain access to the common telephone. Although the telephone corporation giants were our adversary for many years, as told in my book *A Phone of Our Own: The Deaf Insurrection against Ma Bell* (2002), it was, in the present story, the film giants who had far too much control over deaf and hard of hearing people's access to feature movies shown in theaters. And like the story of telephone access, local, state, and national organizations became involved in the activism for access to captioned movies. This was the case when the silent films came out. Soon after this "technological advance," in the 1920s, the soundtrack was added to films and created the first barrier to cultural advancement through films. The introduction of sound films signaled to deaf and hard of hearing people that they were to again be completely isolated from motion picture media—and this lasted for decades. A boon for hearing people, much like the voice telephone, became a bane for deaf citizens.

Closed captioning is not yet a satisfactory solution in theaters as it is with television. Equitable access remains unsatisfactory. I have personally experienced the problems commonly associated with closed captioning theater devices. For example, when I went to a theater with my wife Bonnie to see *The Post*, and again when I saw *Darkest Hour*, I spent frustrating hours adjusting the special glasses so I could adequately follow the dialogue. A slight movement of the head caused the captions to disappear. The glasses also do not fit over regular eyewear. I missed so much of the dialogue that I left the theater disappointed each time. If I were lucky, I might be able to see the film with closed captions on a cable channel months or years later.

I have also wondered why entertainment films were not considered valid learning experiences for K–12 students when the decision was made by the Department of Education to stop funding them. Why would a film like *Darkest Hour*, for example, not be considered educational and be made available to youth through DCMP's streaming service? There is much to learn about Churchill's experience with Germany before the United States became involved in World War II, or about the British Parliament. Another example is *The Post*. The film provides an opportunity to learn about Daniel Ellsberg, freedom of the press, and the Pentagon Papers.

All government-supported film services to deaf adults were stopped at the time entertainment films were no longer distributed; there are no captioned or described film services today comparable to what DCMP offers to educators and students in K–12 programs. In effect, the government assumed that other resources for entertainment films for deaf adults were adequate.

The evolution of the original CFD program through multiple phases over more than six decades has been a remarkable story of the relationship between science, technology, and society. It has also been a story of dynamic educational leaders who effected significant change. The current DCMP, under the auspices of the NAD, is fulfilling well the government goals of providing an exemplary national library of accessible educational media, including DVDs, CD-ROMs, and streaming videos available to teachers and parents/guardians of K–12 students who are deaf, hard of hearing, blind, visually impaired, or deafblind. Its search gateway allows users to selectively search across the DCMP clearinghouse of accessibility-related articles on deafness, blindness, accessibility, captioning, and description, including collaborator websites. The program provides a large collection of free-loan described and captioned educational media, and a clearinghouse of information and research about accessible media.[6]

Evaluation data from users ensure the government that there is great satisfaction that students who are blind, visually impaired, deaf, hard of hearing, or deafblind have the opportunity, through the support from DCMP, to achieve the standards of academic excellence. The program also advocates for equal access to educational media as well as the establishment and maintenance of quality standards for captioning and description by service providers. It is a model serving as a gateway to internet resources related to accessibility.

What began as a government focus on access to captioned films spawned a revolution in the use of educational media in general, and this had a profound impact on the instruction of deaf learners. Funding from the US government always comes with expectations. For sixty years, the captioning program has conscientiously followed mandates defined in its federal authorizing legislation. The mandate to expand media resources far beyond the use of captioned films in the 1960s was an excellent catalyst for enhancing the education of deaf and hard of hearing students. In the new millennium, "described films" added to the program's national captioned film library was a mandate to provide improved access to blind and visually impaired students. Another mandate was to caption films in Spanish. These mandates have resulted in continual improvements in education. DCMP was an important change agent in this regard.

The advances in captioning technology have reaped an explosion of new services and technologies, including closed captioning services (transcription specialists), same-day captioning, closed captioning software, and automatic closed captioning on YouTube. As an active professional for four decades, I have seen the captioning technologies reap great benefits in the academic arena. These and other technologies will certainly shape a new generation of learners.

## THE EXPLOSION OF SOCIAL MEDIA

Over the past decade, social media has become a game-changer for hearing and deaf people alike. The benefits of participating in social media are many. Ryan Commerson, for example, a media strategist for Communication Service for the Deaf, explained that he has used Facebook, Twitter, Instagram, and LinkedIn for social networking, sharing ideas, self-promotion, and "planting ideological seeds for social change."[7] He believes that social media has improved the way people interact, and how they relate to each other.[8] Social media has also "given the Deaf community the tools we need to mobilize, to reconnect, and organize," and it's changed how the Deaf community is perceived by others.[9]

Similarly, Tania Pierre, a deaf writer, claimed that popular social media sites like Facebook and Instagram have opened things up for deaf and hard of hearing people by making it easier to "express themselves with videos, photos, memes, hashtags, gifs, posts, art and the written word—all visual platforms."[10] Through social media, inexpensive communication with hearing family members has been possible, even while living abroad. In isolated areas of the world, such communication has also been affordable and accessible with Skype, Facebook Messenger, and WhatsApp. YouTube and Netflix have made strides in offering closed captioning on all their shows and movies. Pierre writes that, "Emails, social media and apps help me to have a conversation independently, in my own 'voice.'"[11]

Through the years that followed the development of the internet, on an increasing basis social media empowered and connected deaf people and the hearing people with whom they communicated.[12] However, access has remained problematic for deaf users, primarily because of the lack of captioning. As Kožuh and Debevc point out, due to the rapid development of social media, the platforms change regularly, and social media developers, educators, social inclusion advisors, and policy makers need to be advocates for how to apply social media as an inclusive tool for participation in society.[13]

The use of visuals and captions should, in general, make for easy accessibility and enhance how social networks can be increasingly effective with deaf users. But, as Sofia Enamorado summarized in December 2019, social media remains inaccessible for many deaf and blind people. "The overall trends demonstrate that social video is on the rise, but is video accessibility rising with it?"[14] She wrote that the most popular platforms for videos are YouTube, Facebook, and Instagram. "Most social media platforms don't allow users to publish captions or audio descriptions to videos."[15]

Automatic speech recognition, a common method for captioning social video, has been problematic. "The captions are often inconsistent and with poor grammar."[16] Enamorado writes that platform developers are increasingly waking up to the need to add more accessibility features.

Amma Marfo described Twitter's subtitling, which has only recently been added as an accessibility option: "Available in the Twitter Media Studio, the process requires the uploading of a SRT (SubRip Subtitle) file, composed with relative ease via Notepad or Text Edit on your computer."[17] Both YouTube and Facebook came under fire when they first attempted to use an "autocaption" option. Marfo suggested uploading a video and then the SRT file with the video's script and captioning timestamps. For YouTube, she explained that "Users wishing to assure a video's accessibility for [deaf and hard of hearing people] have to revert back to the 'Classic' version of the studio and complete the task there."[18] For LinkedIn, one must upload the video, click the Edit icon for Video Settings, and Select File and upload your SRT file. "Save the file addition, and click 'Post.' From that point forward, your video will appear with subtitles."[19] Many users do not have the technical know-how to do this.

In 2020, Instagram rolled out a captioning option; captions can be enabled by simply selecting that option in the Advanced Settings menu.[20]

As discussed on Twitter, free captioning on social media platforms and video conferencing platforms is critical. In an article in *Hearing Health and Technology Matters*, Gael Hannan wrote that unlike Google Meet or Microsoft Teams, by November 2020, she explained, Zoom did not offer the free ASR option. During the pandemic, deaf people were not asking for special consideration. "We are asking for the same level of communication access that hearing people have . . . We *need* captioning."[21] Hannan explained that not only should Zoom step up, but *all* forms of media should offer text interpretation, *free of charge*.

ASR captions are included in such virtual meeting platforms as Google Meet and Microsoft Teams, and if Zoom adopted it, this might reduce the number of people switching to other platforms.[22] And in a recent interview with National Public Radio, hearing health advocate Shari Eberts argued, "Why should we have to pay to use the feature we require for equal access?"[23] By May 31, 2020, 40,000 signatures had been collected in support of the petition "Provide Free Captions for People with Hearing Loss on Video Conferencing Platforms." Before March, Zoom was beta testing auto captioning that was directly integrated into its platform for paid accounts and was considering making the service free for deaf and hard of hearing people with verification via an audiologist note. However, when the pandemic hit, understandably, "this project moved to the back burner as Zoom focused on beefing up security and infrastructure to support its increased volume."[24]

In the *Washington Post* in December 2020, Eberts explained further that Zoom has the capability to support CART, but "with captioning services running at $200 per hour or higher, this is not a realistic option for many people with hearing loss, especially for personal conversations."[25]

The Association of Late-Deafened Adults tweeted on November 17, 2020 that "Zoom must prioritize auto captioning & commit to making it available to all. If they won't, try Google Meet or MS Teams. Both have free captioning for all. It's clear which platforms care about people with hearing loss."[26]

Regardless of the platforms, there is a slow march toward accessibility in social media, but not quickly enough to satisfy most deaf users. Accessibility should not be an afterthought. "We've just simply gone backwards in accessibility on the platform," says deaf YouTuber Rikki Poynter. "I hate that they listened to hearing creators over the three deaf creators, including myself, that told them they should've kept it."[27] Deaf TikTok creator Chrissy Marshall said, "My experience is TikTok captions are just inconvenient to a lot of people."[28]

Veronica Figueroa Fernandez explains that the biggest drawback for closed captions is "the number of file formats available, and knowing which format is compatible with each platform." Viewers are also dependent on the video publisher to ensure that the closed captions are visible against all backgrounds. And third, some viewers may not have the physical ability to turn captions on or off. With regard to open captions, they are automatically displayed and viewers do not have to worry about how to toggle them on. There is also more versatility in choosing the font color and size when open captions are used.[29]

Even though there are many challenges to providing captions on social media platforms, developers should examine viewer preferences, and brands need to rethink their approach to video marketing.[30] One survey in the United States, for example, reported that 80 percent of consumers are more likely to watch an entire video when captions are available.[31]

## CONCLUSION

At the time I was typing a final draft of this epilogue, I had a personal experience that reinforced the hope that captions in theaters will evolve in the years to follow. I had received an email message from hearing friends, Joan and Dick Haviland, in Honeoye Falls, New York, who wrote, "Check out open caption night at the Cinema - Wednesdays!

We must have a movie date with the 4 of us." This reminded me of the original legislation for captioned films and the hope that it would encourage deaf and hearing moviegoers to share the cultural and entertainment learning experiences through accessible films. My wife Bonnie and I had not done this before. The Havilands are very good friends, but I had simply assumed they would not enjoy a captioned film. I told Joan about working on this history of captioned films and that her note to me was timely. She replied, "I have said to Dick many times that I wished all movies had captions so that deaf people could enjoy them, too. Maybe your book will raise the consciousness of theater owners so that everyone can enjoy going to the movies!! Can't wait to double date!"[32]

Around this same time, I watched a vlog made by Sean Forbes, performance artist and cofounder of the Deaf Professional Arts Network (DPAN). In the video, which was posted on Facebook, Sean was shown leaving a theater with his wife and three-year-old daughter. The family did not look happy.

On the video, Sean signed: "NO CAPTIONS AGAIN! Went to AMC Theatres in Southfield to ring in the new year with my family and watch *Mary Poppins Returns*, which my hearing daughter was so excited to see, but her deaf parents weren't able to."[33] Sean explained that the CaptiView closed captioning system didn't work, and he and his wife had to explain the situation to their daughter, who was very upset that they were leaving after the film had begun.

Importantly, as this book went to press, AMC Theatres, the world's largest movie theater chain, announced that it is adding onscreen captions at hundreds of US locations in an effort to make movie-going more accessible. "Inclusive programming is core to AMC's strategy, and we're proud to lead the theatrical exhibition industry by making some open caption showtimes available at hundreds of our locations nationwide."[34]

Despite this and other wonderful advances in captioning, deaf people still continue to experience frustration in not having the same opportunities afforded hearing moviegoers. The words of Sean Forbes are a powerful call to action: "Everyone should be able to enjoy the movies whenever they want, wherever they want, with whomever they want."[35] This remains an unfulfilled challenge for a new generation, who must continue the work of the many captioning pioneers and advocates to "turn on the words!"

# AFTERWORD

Jason Stark

Thomas A. Edison, credited by Dr. Harry Lang in this book as developing the motion picture, challenged his staff with this slogan: "There's a way to do it better—find it."[1] In the chapters that followed, Harry has also documented the remarkable achievements of numerous individuals who followed the same mantra and found better ways to provide equal access to films and videos for deaf persons.

This has been the guiding principle of what is now known as the Described and Captioned Media Program (DCMP)—revolutionizing techniques for the development of captioned and described educational media, and for its delivery. This book articulates the modest start of Captioned Films for the Deaf, Inc. in Hartford, Connecticut with about thirty 16 mm films. Then, after several reorganizations and name changes, it leads us on the exciting adventure of the DCMP that has evolved into a highly successful federally funded project serving 124,000 members with a collection of over 8,000 accessible educational film titles (including 886 titles in Spanish). The most modern technologies have brought the project to this point, with over 95 percent of all DCMP user transactions being streamed media. There is now virtually no limit to the number of users who can utilize accessible educational media from the project.

Since 1991, seeking ways to "do it better" has resulted in a remarkable number of innovative accomplishments for the project, as documented in this book, and these include, to name a few: (1) the first-ever accessibility guidelines for both educational captioning and description (for blind and visually impaired students); (2) the first-ever streamed captioned (and later described) educational media; and (3) the first-ever fully accessible bilingual (English and Spanish) educational media.

An innovative spirit clearly was also possessed by many of the program pioneers, including Dr. Malcolm Norwood, who was a huge admirer of Edison. I suspect Mac knew the following words were uttered by the inventor to a beginning staffer who inquired about research laboratory rules: "Hell there are no rules here. We are trying to accomplish something."[2]

For it was in 1980 at a national conference on captioning held at the National Technical Institute for the Deaf that Mac reminded the audience how in the beginning there had been very little research done on the effectiveness of captioning as a means of providing verbal information to deaf persons. There were no specific rules or guidelines, and it was simply assumed that captions would be a valid alternative to the soundtrack. Mac smilingly referred to how, similar to Edison, he and others learned to "fly by the seat of your pants."[3] But, he indicated that everything done was based on what they knew the kids could do (later research proved the approaches to be correct). Mac noted also that Confucius once said: "The Cautious seldom err." When rules don't exist, dedicated, knowledgeable, and cautious experts have to create them.[4]

In asking me to write this afterword, Harry, in effect, requested that I develop the outline for my own "dream book," similar to the one developed by John Gough in 1960 and briefly described elsewhere in this book. Over a half century ago Gough correctly predicted that a much broader media service would supersede CFD.[5]

Similarly, I predict that DCMP will continue to play a critical role, along with other organizations, in leading the country's efforts in accessible media research and development. I am confident in this prediction as long as the government continues its long-held approach of generally selecting nonprofit sector entities (e.g., NAD, CEASD, AED, NFB) to provide leadership and administration for the project. The US Department of Education's commitment to teaching and learning, and to nonprofit (consumer) knowledge of the needs of people with disabilities, will have an impact beyond current specific project objectives. Future leaders must, and will, develop new rules and guidelines for access when they don't exist.

I've witnessed three major media format shifts during my career: 16 mm to VHS, VHS to DVD, and DVD to streaming. Each of these introduced a new era requiring that the project transition to the newer methods of generating media access. Of course, new technologies will continue to be invented, and I predict that DCMP will continue to be wildly innovative in its development of technology to increase sensory impaired access. For example, virtual, mixed, and augmented reality (sometimes referred to as XR, X reality, or cross reality) content and 360-degree videos are finding their way into classrooms and will need to be made accessible.

In addition, DCMP will remain on the forefront in the utilization and development of technology to produce accessibility more efficiently. Text-to-speech engines are improving and their use to provide descriptions will become a more user-accepted alternative to the currently preferred human-recorded dialogue tracks. Also, automatic speech recognition coupled with machine learning techniques will continue to improve and be a viable tool to generating high-quality captions.

Until educational content creators see the market demand for producing media with descriptions, it simply will not happen. As such, I believe the DCMP should continue as the leading source of free, fully accessible educational media for the foreseeable future. The activities of content distribution are important and the project will continue to adapt new platforms as they are invented (such as the Roku and AppleTV channels). Further, DCMP will remain committed to following new trends in educational media such as short-form content on platforms like YouTube. In 2019, DCMP reached 2 million total views on its YouTube channel.

And, as my father, Bill Stark, once said during a newspaper interview about the impact of technology, we will continue to develop the internet so that one day it will be as friendly to deaf and hard of hearing people as TV is now.[6]

In addition to being the premier source of fully accessible media, DCMP has had a long history of providing extraordinary assistance to teachers in the classroom (and to families) in how to use media to improve results for students with disabilities. This work began decades ago with the generation of lesson guides and most recently includes the correlation of the media to state and national curriculum standards. Future innovations will build on the student account feature and allow for the creation of interactive lessons. These lessons can include videos, clips, and questions, and reinforce educational content by requiring the student to rewatch content if questions are not answered correctly.

Additional items in my "dream book" involve DCMP being a tool to support interactive approaches to learning. It is true that learning is a function of interaction, but video platforms present teachers with the chance to increase engagement of students and script levels of interactivity previously unavailable.

My first dream is drawn from Dr. Lang's own research with his colleagues at the National Technical Institute for the Deaf, called "Structured Online Academic Reading," in which students could click on terms in the online text and get definitions, explanations, in sign or text, photos, or links for further reading on the web. Using DCMP's existing interactive transcript feature, links to additional content could easily be added. These could be resources curated by DCMP's educational staff, or perhaps be added by a teacher or parent for use in the classroom or home environments.

Students sharing ideas by developing lessons of their own can make the dream of using technology to boost students' communication skills a reality. When students take ownership of their own learning they become more engaged. DCMP could provide existing clips, playlists, and content creation tools to allow students to create lessons and/or projects and connect with other classes, either in the same school or different schools.

Teachers will have greater flexibility with teaching methods, as lessons are created that consist of mixed media, gamified elements, videos, video clips, and more. Old and new resources can be combined into unique lessons plans that meet the special needs of the deaf and blind population. DCMP's upcoming lesson builder tool focuses entirely on DCMP content. With the proliferation of educational content available from other sources, DCMP could allow teachers to build interactive lessons, including self-created videos (assuming they are accessible), graphics, content slides, and other external resources.

As was written over a decade ago in a historical overview of the project posted on the DCMP website: " . . . the story of captioning does not end at this point, because it is still being written by advocates."[7] This remains true today and certainly applies to description as well. Description for those who are blind and visually impaired is a relatively new field, and I cannot wait for breakthroughs that will occur over the next decades. There will be many new advocates, including men and women who benefited by learning through captions and descriptions, who will find ways to "do it better" and write new chapters in this fascinating story of access.

## ACKNOWLEDGMENTS

I dedicate this book to the individuals who led the movement to bring access to captioned films to Deaf people through the past six decades. Many of these pioneers were associated with the CFD, a project begun in 1958 and sponsored by the US Department of Education. John A. Gough, the first "chief," was at the helm for the first decade as he laid the foundation for the subsequent half century of remarkable progress. The skills Gough previously demonstrated at the Oklahoma School for the Deaf illustrate well his remarkable ability to inspire collaboration in accomplishing the goals he set forth. After the bombing of Pearl Harbor in December 1941, the country mobilized for war. With prices frozen on foreign products and shortages of coffee, meat, rubber, gasoline, and other staples, many employees flocked to the factories to support the government. In a tribute to their former superintendent, John A. Gough, the alumni reminisced about how he had asked the students for teamwork in maintaining the dormitories and campus, pitching in to rake leaves, planting potatoes, ironing clothes, doing laundry, and much more as they helped maintain the school's operation during the war. "Do you look nostalgically at the Supt. John A. Gough era as the good old days?" asked Carol Eileen (Evanhoe) Sponable, a profoundly deaf woman who was at the school at that time. "I do. It is because we formed an extraordinary teamwork, resulting in lifetime bondings and friendships."[1]

In many ways these notions of teamwork, lifetime bonding, and friendships characterize the first decade of the history of CFD. The respect Gough's colleagues had for him as the leader who initiated the federally funded program was everywhere apparent. The same camaraderie continued among the staff in this program throughout its history, as well as those in programs joining forces with CFD to bring captions to deaf and hard of hearing people everywhere. I also personally experienced the camaraderie among those many people who were so willing to provide me with information and anecdotes for this book.

They included attorneys, accessibility advocates, staff at NCI, and deaf consumers, to name a few.

In 1960, at a conference to kick off the plans for implementing CFD, among the many recommendations made by the conferees was that the new director, John A. Gough, arrange for preparation of a history of the program. Fifty-eight years later, I was honored to be invited by four of the pioneers—Gil Delgado, Bill Stark, Doin Hicks, and Norb Nathanson—to write the history in detail. I greatly appreciate the insights and suggestions these four gentlemen offered over the past few years as this story took form. They also helped to find some funds to support the research and writing. Special appreciation is due Jason Stark and the NAD for this financial assistance.

The autobiographical notes of Gough and the effort to document the first decade of CFD's history by James J. Kundert, a staff member under Gough, were recently found by Susan Gough Updegraff, John's daughter, in boxes stashed away for more than half a century. The writings provided additional perspectives that reveal the challenges these leaders faced in creating and disseminating new technology to enhance access through captioning for deaf people of all ages.

Under Gough's supervision, Gilbert L. Delgado came on board to help expand CFD's role in the media-based instruction revolution that was taking place, and to coordinate the agenda of research studies begun in the late 1960s. In 2014, Gil independently revived the idea of documenting the program's history, working closely with Bill Stark, Doin Hicks, and Norb Nathanson to formulate the outline and gather together some documents. Without Gil's initiative and his invitation to me to author this book, it may not have been pursued for many more years.

In 2019, Gallaudet University Press asked me to expand the story to include others who contributed meaningfully to the technological breakthroughs and the efforts to bring more universal access to the spoken word. At this time, the book became a first attempt to document the broader history of captioning, although films remained the primary focus. As a result, the story of access to films encompasses theaters, television, streaming services, the internet, and other technologies.

Malcolm J. Norwood, the "Father of Closed Captioning," was hired by Gough to foster national interest in accessibility through captioning, and, subsequently, Norwood became CFD chief after Delgado took a position at Gallaudet College (now University).

I appreciate the support of Len Novick and H. Joseph "Joe" Domich Jr. who directed film distribution and managed other aspects of CFD for nineteen years. Bill Stark, nicknamed "The Language Man,"[2] was my primary advisor while I was writing this book.

Bill's son, Jason Stark, known affectionately as the "Wizard," began managing the program in 2011. Bill and Jason Stark provided me with numerous data, anecdotes, and perspectives related to the last three decades of the program. As leaders for more than half of the program's existence, the Starks (and I include Bill's wife, Dianne, a master teacher of deaf students who performed exemplary work in developing the *Captioning Key*) provided a successful resource program, the Described and Captioned Media Program (DCMP), that has been widely praised by teachers, parents, vendors, and various audiences and collaborators involved in the education of deaf, blind, and deafblind learners. The resources are under the auspices of the NAD and funded by the US Department of Education, whose foresight and support has made it possible to bring the power of the spoken word to people who have long been denied access.

The DCMP website includes more than 300 articles about captioning, deafness, blindness, and education. I owe much gratitude to the many authors of those articles and to Bill Stark for organizing these resources, writing or editing many of them, and posting them for the benefit of consumers and scholars. I have paraphrased some of the information I obtained from these articles and cited the original authors. Bill's writings, especially, and his emailing with me have considerably enriched this book with his own writing voice.

At one point while doing the research, I received a supportive email from a friend, Bob Davila, former deaf director of NTID and former president of Gallaudet University. "You are correct," he wrote, "in stating that little of this effort by the government to create a bridge to the world of speech for people has been properly recorded and preserved. Many of the people who were involved in developing captioning services are no longer alive and there are not many sources of information or records."[3] Sadly, two of the four people who had originally invited me to author this book, Gil Delgado and Norb Nathanson, passed away while I was completing the manuscript and would not see the final published work.

Bob Davila's note of caution about the availability of information was a point well taken. But, I soon found the archives at Gallaudet University, RIT, and other libraries invaluable. In this regard, I especially thank Michael Olson and Christopher Shea at Gallaudet University, and Joan Naturale at NTID at RIT. Morna Hildebrand located articles and photographs at RIT as well. Vincent Baiera assisted with the production of several photographs. Appreciation is also due David H. Pierce for sharing articles from his own archives as well as his perspectives. I am grateful also to George Potanovic, who provided a treasure of information on deaf film maker Ernest Marshall.

Communication with friends and colleagues through social media about their personal experiences with captioning included names too numerous to list here. Some of their names are provided throughout this volume; others are found in the endnotes.

Susan Gough Updegraff provided reminiscences about her father, John A. Gough. As mentioned earlier, Susan found some historical notes for the first six years of the program that had been written by James J. Kundert, a deaf man hired by Gough in the early 1960s. Kundert's notes were invaluable in providing more details on the challenges faced by the program's first leader.

Toni Atencio deserves my appreciation for her assistance and support to Gil Delgado and me as we communicated for two years while the book was developing. Zainab Alkebsi and Sheila Conlon-Mentkowski assisted with information related to the battle for open captions in theaters. Nanci Linke-Ellis, general manager of Captionfish, shared her many years of experience in working with studios to bring captions to deaf and hard of hearing moviegoers. Karen Peltz Strauss was especially helpful in that regard, pointing out important issues she faced as a leading telecommunications policy advocate. Called a "legend" when she retired from the FCC, she was extremely enthusiastic about helping me with issues of accessibility and television. Karen also reviewed the manuscript with many helpful suggestions, as did Jim House, Carl Jensema, Joe Domich, Len Novick, and Ernie Hairston.

I am grateful to Toby R. Silver for providing information about her work with the Movie Access Coalition and Television For All, and

John Waldo, an attorney, who shared his knowledge and experience with the issue of open captions in theaters. John Koskinen, who served for twelve years as an organizer and chairman of the board of the National Captioning Institute, and Jeff Hutchins, one of the first employees of The Caption Center at WGBH in Boston and the first executive in charge of technology development, at the National Captioning Institute (NCI), shared invaluable information about captioning and NCI.[4] Thanks also to Robin Pierce for information on The Caption Club.

Ivey Pittle Wallace, Katie Lee, Angela Leppig, Deirdre Mullervy, and Valencia Simmons at Gallaudet University Press have taken this book to completion with excellence, including editing, advising, working with copy editors, inserting photographs and graphics, planning for marketing, and a variety of other work. I offer very special thanks to Katie for her patience and wonderful support as I completed this book during the pandemic.

And on behalf of deaf people everywhere, I thank the many thousands of people who worked as teachers, captioners, inventors, and other crusaders to bring us access to the spoken word in films.

To my wife, Bonnie, for her endless loving support. I quote the great William Shakespeare—"I would not wish any companion in the world but you."

# NOTES

## Preface

1. Jim House, personal communication to the author, November 3, 2019.
2. Frank B. Withrow letter to John A. Gough, June 11, 1992. Letter shared with author by Susan Gough Updegraff, January 2019.

## Introduction

1. John A. Gough, "Captioned Films for the Deaf," *School Life* (January–February 1962): 26.
2. I. King Jordan, "Opening Remarks," *Written Proceedings of the National Conference for Closed Captioning of Local News,* (November 21–23, 1991), 15. "Is it CC?" *Andrea's Buzzing About . . .* (blog), June 7, 2008, https://qw88nb88.wordpress.com/2007/06/08/is-it-cc/.
3. Peggy Cagle, email to the author, December 22, 2018.
4. Battinto Batts Jr., "Getting the Message: Technology Captures Hopes of Deaf People," *South Florida Sun Sentinel* (Fort Lauderdale), January 9, 1995, 11.
5. Malcolm J. Norwood, "Captioned Films for the Deaf," *Exceptional Children* 43, no. 3 (1976). Available from https://dcmp.org/learn/84-captioned-films-for-the-deaf.
6. Sara Nović, "The Frustrating Fight for Closed Captions on In-Flight Entertainment," *Condé Nast Traveler*, January 16, 2020, https://www.cntraveler.com/story/the-frustrating-fight-for-closed-captions-on-in-flight-entertainment.
7. Nović, "The Frustrating Fight."
8. Nović, "The Frustrating Fight."

## Chapter 1

1. James H. Logan, "Books, Pictures and Illustrative Apparatus for the Deaf and Dumb," *American Annals of the Deaf and Dumb* 15 (1870): 93–104.
2. Logan, "Books, Pictures," 103–4.
3. John Fullerton and Astrid Söderbergh Widding, *Moving Images: From Edison to the Webcam* (United Kingdom: John Libbey Publishing Ltd., 2016), 6.
4. John Columbus, "A Sketch of Thomas Edison and the Development of the Motion Picture," *The River,* March 9, 2012, 3, https://ufdcimages.uflib.ufl.edu/UF/00/10/13/63/00114/03-09-2012.pdf.
5. The cause of his deafness has been attributed to a bout of scarlet fever during childhood and recurring untreated middle-ear infections. Around the middle of his career, Edison attributed his deafness to being struck on the ears by a train conductor as a young man when his makeshift chemical laboratory in a boxcar caught fire and he was thrown off the train in Smiths Creek, Michigan, along with his apparatus and

chemicals. In his later years, he modified the story to say the injury occurred when he was twelve years old and a conductor, in helping him onto a moving train, lifted him by the ears.

6. Thomas A. Edison, *Menlo Park: The Early Years, April 1876–December 1877*. Vol. 3, The Papers of Thomas A Edison, ed. Robert A. Rosenberg, Paul B. Israel, Keith Nier, and Martha J. King (Baltimore: Johns Hopkins University Press, 1989), doc. 1117.
7. Columbus, "A Sketch," 3.
8. "Thomas Edison," *Science and History* (blog), http://science-history.blogspot.com/p/thomas-edison.html.
9. "Origins of Motion Pictures," *Library of Congress*, https://www.loc.gov/collections/edison-company-motion-pictures-and-sound-recordings/articles-and-essays/history-of-edison-motion-pictures/origins-of-motion-pictures/.
10. "Silent Film Collection," *Carl Bennett's File Cabinet*, http://www.silentera.com/family/carl/films/silentsCollection.html.
11. Patt Lee Garland, "The Silent Era," *Cinemashares*, https://www.cinemashares.com/hollywoodhistory/CS_silentera_ch1.html.
12. Mildred Albronda, *The Magic Lantern Man: Theophilus Hope d'Estrella* (Fremont: California School for the Deaf, 1985).
13. "Will Play Ball to Buy Picture Machine," *Daily Arkansas Gazette* (Little Rock, Arkansas), May 5, 1912, 40.
14. John S. Schuchman, *Hollywood Speaks: Deafness and the Film Entertainment Industry* (Urbana: University of Illinois Press, 1988), 45.
15. The Indiana State School for the Deaf, *Eighty-sixth Annual Report*, September 30, 1929.
16. Indiana State School for the Deaf, *Eighty-sixth Annual Report*.
17. Schuchman, *Hollywood Speaks*, 21.
18. Schuchman, *Hollywood Speaks*.
19. Schuchman, *Hollywood Speaks*.
20. Schuchman, *Hollywood Speaks*.
21. "Visionary Leader (July 2014): George Veditz," *Gallaudet University*, https://www.gallaudet.edu/about/history-and-traditions/george-veditz.
22. John S. Schuchman, "The Silent Film Era: Silent Films, NAD Films, and the Deaf Community's Response," *Sign Language Studies* 4 (Spring 2004): 232.
23. *The Deaf-Mutes' Journal* 58, no. 3 (January 17, 1929): 2. https://islandora.wrlc.org/islandora/object/dmjournal%3A239/datastream/PDF/view.
24. *The Deaf-Mutes' Journal*, 2.
25. "Deaf Protest Talkies With No Captions," *Hartford Courant* (Hartford, Connecticut), April 3, 1929, 5.
26. "Deaf Protest," 5.
27. "Talkie Captions Asked by Deaf: Hard of Hearing Have Difficulties in Following the Plot," *The Journal Times* (Racine, Wisconsin), June 14, 1929.
28. "Sound that Shook Hollywood," *New York Times*, September 1977.
29. "Talkie Captions Asked by Deaf," 35.
30. *The Deaf-Mutes' Journal* 58, no. 3 (January 17, 1929): 2.
31. "Talkie Captions Asked by Deaf," 35.
32. "Talkie Captions Asked by Deaf," 35.

33. *The Deaf-Mutes' Journal* 58, no. 3 (January 17, 1929): 2.
34. *The Deaf-Mutes' Journal* 58, no. 3 (January 17, 1929): 2.
35. "Living with a Genius," *Reader's Digest,* March 1930: 1042–44.
36. Trent Batson and Eugene Bergman, eds. *Angels and Outcasts: An Anthology of Deaf Characters in Literature* (Washington, DC: Gallaudet University Press, 1985), 303.
37. Schuchman, *Hollywood Speaks.*
38. Ladner, "Silent Talkies," 323–25.

CHAPTER 2

1. Harry G. Lang, *A Phone of Our Own: The Deaf Insurrection against Ma Bell* (Washington, DC: Gallaudet University Press, 2002); and Karen Peltz Strauss, *A New Civil Right: Telecommunications Equality for Deaf and Hard of Hearing Americans* (Washington, DC: Gallaudet University Press, 2006).
2. Sylvia Sanders, "The Use of Visual Aids in Teaching the Deaf" MA thesis, (Gallaudet College, 1946), 21.
3. Sanders, "Visual Aids," 21.
4. Romero, "Emerson Romero."
5. D. Bangs, "Deaf Access Programming," in *Gallaudet Encyclopedia of Deaf People and Deafness, Vol. 3*, ed. J. V. Van Cleve (New York: McGraw-Hill, 1987), 270–72.
6. *The Cavalier*, October 1947. Rochester Institute of Technology Archives.
7. Edmund B. Boatner, "Captioned Films for the Deaf," 1980, https://dcmp.org/learn/static-assets/nadh93.pdf.
8. Emerson Romero, "The National Film Library for the Deaf," *The Silent Worker*, January 1951, 23. https://archive.org/details/TheSilentWorker0305January1951/page/n22/mode/1up.
9. Bill Stark, "How Bird Hunting in North Carolina Saved Captioning," *Described and Captioned Media Program,* https://dcmp.org/learn/36-how-bird-hunting-in-north-carolina-saved-captioning.
10. Upon graduation from the New York School for the Deaf at Fanwood in 1931, Marshall became a projectionist, and from 1931 to 1994 he was a skilled and dedicated projectionist for New York–area Deaf clubs and organizations. During that time, he estimated having arranged more than 2,042 programs, including more than 500 films at the Union League and another 500 at the Long Island Club for the Deaf. In 1936, he found a job at a Manhattan hotel and purchased a 16 mm movie camera to pursue his "self-taught and self-financed" career in filmmaking. On May 8, 1937, at St. Ann's Church for the Deaf, he released a feature film in ASL, *It Is Too Late*, the first of eleven produced during his career—seven feature films and four short films. In the late 1930s he began to produce entertainment films that emphasized "stories." His association with deaf stage actors (including Emerson Romero) convinced him that the most important element of filmmaking is the story. This, of course, was precisely what most audiences wanted—a good story. See Schuchman, *Hollywood Speaks*, 45.
11. George Potanovic Jr., email to author, March 21, 2019. Much of the information about Ernest Marshall was contributed by George Potanovic Jr.
12. George Potanovic Jr., email to author.
13. Ross Hamilton to John K. Norton, May 21, 1947, quoted in James J. Kundert, "Cap-

tioned Films for the Deaf" (unpublished manuscript, 14–15). Courtesy of Susan Gough Updegraff.

14. Kundert, "Captioned Films for the Deaf," 14–15.
15. Kundert, "Captioned Films for the Deaf," 15.
16. Kundert, "Captioned Films for the Deaf," 15.
17. "Talkies for the Deaf," *The Sunday Herald* (New South Wales), May 14, 1950.
18. Edmund B. Boatner, "Captioned Films for the Deaf," 1980, https://dcmp.org/learn/static-assets/nadh93.pdf.
19. Boatner, "Captioned Films."
20. Boatner, "Captioned Films."

### Chapter 3

1. Edmund B. Boatner, "Report of the Progress of the Work of the Committee on Captioned Films. Addendum to Minutes of the Twenty-fifth Regular Meeting of the Conference of Executives of American Schools for the Deaf, June 29, 1953." Gallaudet University Archives.
2. Rakow was later made supervising teacher of the vocational department of the American School, a position he held until his retirement in 1963.
3. John A. Gough, "'Talkies' for the Deaf Through Captioned Films," *Educational Screen & Audio-Visual Guide* 40 (May 1961).
4. The Conference of Executives of American Schools for the Deaf, the Convention of the American Instructors of the Deaf, the Alexander Graham Bell Association, the NAD, and the National Fraternal Society of the Deaf all supported the proposal.
5. James J. Kundert, "Captioned Films for the Deaf" (unpublished manuscript, 49). Courtesy of Susan Gough Updegraff.
6. Kundert, "Captioned Films for the Deaf."
7. Enis Mulolli, "Which are the most frequent problems faced in initial phase by start-ups?" question posted on *ResearchGate*, https://www.researchgate.net/post/Which_are_the_most_frequent_problems_faced_in_initial_phase_by_start-ups

### Chapter 4

1. Duane Gerstenberger, "Library Services for the Blind: A Brief Review and Overview." *Future Reflections* 4, no. 2 (April/May/June 1985), https://www.nfb.org/images/nfb/publications/fr/fr04/issue2/f040202.html
2. National Library Service for the Blind and Physically Handicapped, Library of Congress, "History," https://www.loc.gov/nls/about/organization/history/.
3. National Library Service, "History."
4. President Theodore Roosevelt, letter to the secretary of the Treasury, September 19, 1935.
5. John Cory, Chief, Circulation Department, New York Public Library. In Patricia Cory, "Report of a Conference on the Utilization of Captioned Films for the Deaf Held at the Lexington School for the Deaf, New York, NY, June 8–10, 1960." US Department of Health Education, and Welfare.
6. Boatner, "Captioned Films."
7. Verner W. Clapp, "Some Problems in Library Service for the Blind" (paper pre-

sented at the nineteenth biennial convention of the American Association of Workers for the Blind, Indianapolis, Indiana, July 7–11, 1941), in *Proceedings of the Nineteenth Biennial Convention of the American Association of Workers for the Blind* (n.p., n.d.).

8. "Captioned Film Library Bill Approved by 85th Congress," *The Pelican* 79, no. 4 (January 1959): 1–2. *The Pelican* was a newsletter published by the Louisiana State School for the Deaf.
9. Boatner, "Captioned Films," 9.
10. Boatner, "Captioned Films," 9.
11. Kundert, "Captioned Films for the Deaf."
12. Boatner later published an article in the *American Annals of the Deaf* (96, no. 3, 346–352) about a story he had learned years later. He described how Graham H. Anthony helped rescue the bill by flying to Washington when he heard about the delay.
13. Boatner, 1980. Although the bill passed in 1958, it wasn't until 1967 that Anthony relayed the full story of how the bill reached the House floor.
14. Catalog of Captioned Films/Videos for the Deaf Films (1991). St. Petersburg, FL: Modern Talking Picture Service, Captioned Films/Videos for the Deaf.
15. Original text of PL 85-905, the public law that originally established Captioned Films for the Deaf. Also known as the Captioned Films Act of 1958. September 2, 1959. https://dcmp.org/learn/212-public-law-85-905.
16. After this law was passed, CFD, Inc. became CFD.
17. Boatner, "Captioned Films."

### Chapter 5

1. Oscar M. Pittenger, *Indiana School for the Deaf Eighty-Third Annual Report* (Indianapolis: Indiana School for the Deaf, 1927).
2. Patricia Cory, *Report of a Conference on the Utilization of Captioned Films for the Deaf Held at the Lexington School for the Deaf, New York New York, June 8-10, 1960* (Washington, DC: US Department of Health Education, and Welfare, 1960).
3. Bill Stark, "The Logic of the Motion Picture in the Classroom: Films in Schools for the Deaf (1915–1965)." https://dcmp.org/learn/254-the-logic-of-the-motion-picture-in-the-classroom-films-in-schools-for-the-deaf-1915%E2%80%931965.
4. "Film Fare," *The Silent Worker*, November 1960, 24.
5. "Destry Rides Again," *The Silent Worker*, December 1961, 4.
6. In 1971, Terry O'Rourke, instructor at the Kendall School at Gallaudet College, joined the group captioning feature films.
7. "Film Fare," *The Silent Worker*, May 1961.
8. Kundert, "Captioned Films for the Deaf," 61.
9. Those ten deaf professionals included Max Friedman, writer; Leon Auerbach of Gallaudet College; Mervin D. Garretson, supervising teacher of the Montana School; Jess Smith, first vice president of the NAD; and Boyce R. Williams, consultant on the deaf and hard of hearing in the Office of Vocational Rehabilitation. Others taking part included Leonard M. Elstad, president of Gallaudet College; Powrie V. Doctor, editor of the *American Annals of the Deaf*; Marshall S. Hester, superintendent of the New Mexico School and president of CAID; Edmund B. Boatner, William J.

McClure, Clarence D. O'Connor, and Stanley Roth, superintendents respectively, of the American, Indiana, Lexington, and Kansas Schools. Then, along with several other representatives of the Office of Education, Gough attended a third conference, this one with the nine-member producing-distributing companies in the United States to discuss obtaining feature films suitable to the program. No date for this meeting was found. It was probably in early 1960.

10. Max Friedman, "C.F.P. Act Passed by Congress," *Empire State News*, March 1960, 23.
11. "John A. Gough, Educator Extraordinary," *The Deaf American*, February 1969, 3.
12. Kundert, "Captioned Films," 62.
13. "Captioned Films Very Good," *The Kentucky Standard*, January 28, 1960.
14. Clipping from Ohio School for the Deaf, February 1961, Gallaudet University Archives.
15. Vinita Seward, letter to John Gough, n.d., Gallaudet University Archives.
16. Participants included Edmund B. Boatner, superintendent and principal, American School for the Deaf, West Hartford, CT; Charles Brooks, New York School for the Deaf, White Plains, NY; Julien Bryan, director, International Film Foundation, New York, NY; Byron B. Burnes, president, National Association of the Deaf, Berkeley, CA; Anita A. Carpenter, Captioned Films for the Deaf, US Office of Education, Washington, DC; Daniel T. Cloud, superintendent, New York School for the Deaf, White Plains, NY; Leo E. Connor, assistant superintendent, Lexington School for the Deaf, New York, NY; John M. Cory, chief, Circulation Department, New York Public Library, New York, NY; Patricia Cory, librarian, Lexington School for the Deaf, New York, NY (Conference Coordinator); Powrie V. Doctor, editor, *American Annals of the Deaf,* Gallaudet College, Washington, DC; Max Friedman, National Association of the Deaf, New York, NY; John A. Gough, specialist, Captioned Films for the Deaf, US Office of Education, Washington, DC; Roy M. Hall, assistant commissioner for research, US Office of Education, Washington, DC; Beatrice Hodgkins, in charge of after school activities, Lexington School for the Deaf, New York, NY; Emily Jones, executive secretary, Educational Film Library Association, New York, NY; Jeannette Johnson, executive secretary, Alexander Graham Bell Association for the Deaf, Washington, DC; Dr. Marcus L. Kenner, New York, NY; Mrs. Dorothy Kraft, Lexington School for the Deaf, New York, NY; William J. McClure, superintendent, Indiana School for the Deaf, Indianapolis, IN; J. Stanley McIntosh, director, Educational Services, Motion Picture Association of America, New York, NY; Kathleen McKinnon, Captioned Films for the Deaf, US Office of Education, Washington, DC; Elliott Newcomb, Encyclopedia Britannica Films, Inc., Wilmette, IL; Alan Morrison, president, Industrial Audio Visual Association, Film Department, Socony Mobile Oil Co., New York, NY; Clarence D. O'Connor, superintendent, Lexington School for the Deaf, New York, NY; Beatrice Ostern, assistant principal, Lexington School for the Deaf, New York, NY; Frances Phillips, principal, Kendall School, Gallaudet College, Washington, DC; J. Pierre Rakow, principal, Vocational Department, American School for the Deaf, West Hartford, CT; Paul C. Reed, director, Instructional Materials, Rochester Board of Education, Rochester, NY; Seerley Reid, assistant director, Educational Media branch, US Office of Education, Washington, DC; Dr. Paul Rotter, assistant to the superintendent, Lexington School for the Deaf, New York, NY; Hugo Schunhoff, superintendent, West Virginia Schools for the Deaf and

Blind, Romney, WV; Roy M. Stelle, superintendent, Colorado School for the Deaf and Blind, Colorado Springs, CO; Jack Tarins, New York School for the Deaf, White Plains, NY; Elizabeth Titsworth, assistant superintendent, New Jersey School for the Deaf, West Trenton, NJ; H.L. Vanderford, film production manager, American Telephone & Telegraph Co., New York, NY; Eleanor Vorce, principal, Lexington School for the Deaf, New York, NY; Boyce R. Williams, consultant, Deaf and Hard of Hearing, Office of Vocational Rehabilitation, Washington, DC.

17. Cory, "Report of a Conference."
18. In consideration of age levels and groups with reference to availability of funds, the conference participants recommended that initial priorities be considered in this order: 40 percent of the available monies to formal groups such as adult clubs, associations, and leagues of the hard of hearing (given first consideration); and then 40 percent to in-school groups such as residential schools, day schools and classes, and speech and hearing centers (second consideration). Religious groups such as the Newman Club, and Sunday schools for the deaf, should be given third consideration (10 percent) and out-of-school groups, including homes, hospitals, camps, and informal groups, should be given fourth consideration (10 percent). Provision should be made for evaluation of captioned films in the collection, by those who use them, as a guide for future selections.
19. John Gough, "Captioned Films for the Deaf," *The Silent Worker*, August 1960, 20.
20. Doin Hicks, "Captioned Films for the Deaf Curriculum Workshops," https://dcmp.org/learn/376-captioned-films-for-the-deaf-curriculum-workshops.

## Chapter 6

1. Other bookings included adult Deaf clubs and organizations, and church groups. "Captioned Films for the Deaf," *School Life* (January–February 1962): 26–27.
2. "Film Fare," *The Silent Worker*, December 1960, 28.
3. Deafened at the age of five from back-to-back illnesses of measles and scarlet fever, Norwood was a graduate of the American School for the Deaf in West Hartford. He had completed his baccalaureate studies at Gallaudet College in 1949 and was editor-in-chief of the college newspaper (*The Buff and Blue*) and captain of the cross-country team. Mac began teaching at the Texas School for the Deaf in Austin (1949–1950), taught at the American School for the Deaf (1952–1954), and served as director of curriculum and supervising teacher in the Intermediate and Advanced Department at West Virginia School for the Deaf in Romney (1954–1960). He married a fellow New Englander, Marjorie Hale, in 1952 and earned his Master's degree in Education from the University of Hartford in 1957. He later earned his PhD at the University of Maryland.
4. "Rockets: How They Work," *Described and Captioned Media Program*, https://dcmp.org/learn/246-rockets-how-they-work; https://dcmp.org/media/6395-rockets-how-they-work.
5. Boyce Williams began his thirty-eight–year career with the federal Office of Vocational Rehabilitation (renamed the Rehabilitation Services Administration in 1967). He ultimately became director of the Deafness and Communicative Disorders branch.
6. Kundert, "Captioned Films," 128.
7. Rodney Wilson Walker, *My Life Story*, 2006. As mentioned in "United Utah Organizations of the Deaf," written and compiled by Jodi B. Kinner and edited by Valerie G. Kinney, 2012, https://www.utahdeafhistory.com/uploads/5/4/2/6/5426987/united_utah_organizations_deaf.pdf.

8. John A. Gough, "Captioned Films for the Deaf:" Brief, *Remarks Prepared for the Hearing before a Special Subcommittee of the Committee on Labor and Public Welfare, United States Senate on S. 2511, a Bill to Provide for an Increased Program of Captioned Films for the Deaf*, August 7, 1962, *Described and Captioned Media Program*, https://dcmp.org/learn/94-captioned-films-for-the-deaf-brief.
9. "Film Fare," *The Silent Worker*, February 1963.
10. "Film Fare," February 1963.
11. John A. Gough, "Captioned Films for the Deaf," *School Life* (January–February 1962): 26–27.
12. "First Educational Films Captioned for Deaf," *News and Reports*, OE34002-4, March 1961, Office of Education, US Department of Health, Education, and Welfare. Reprinted in *The Maryland Bulletin* 81, no. 8 (April 1961): 117. http://msa.maryland.gov/megafile/msa/speccol/sc5300/sc5339/000113/011000/011220/unrestricted/20090109e-013.pdf.
13. "First Educational Films," 117.
14. Robert H. Finch, Secretary, US Department of Health, Education, and Welfare, "Preparation of Professional Personnel," *A Review of Selected Program Activities in the Education of the Deaf by the US Department of Health, Education, and Welfare*, February 1969, 6–7.
15. "Film Fare," *The Silent Worker*, October 1964.
16. John A. Gough, "Captioned Films for the Deaf, Report to Conference of Executives," Riverside, California, April 12–17, 1964.
17. Bill Stark, "Described and Captioned Media Program: Seventy Years of Progress," *Described and Captioned Media Program*, https://dcmp.org/learn/379-described-and-captioned-media-program-seventy-years-of-progress
18. John A. Gough, "Gullible's Travels" (unpublished personal memoirs), 96–99. Courtesy of Susan Gough Updegraff.
19. Gough, "Gullible's Travels," 97.
20. Gough, "Gullible's Travels," 98.
21. Robert R. Davila, email to author, August 2, 2019.
22. Bill Stark, email to author, August 4, 2019.
23. "Help Center: Browse by Standards," *Described and Captioned Media Program*, https://dcmp.org/learn/588-browse-by-standards.
24. Bill Stark, "A Video Decalogue…," *Described and Captioned Media Program*, https://dcmp.org/search?q=decalogue&t=l.
25. Stark, "A Video Decalogue."
26. Bill Stark, email to author, August 4, 2019.
27. Robert Baughman, letter to Captioned Films for the Deaf, Gallaudet University Archives.
28. *The Ohio Chronicle*, February 10, 1962. Gallaudet University Archives.
29. "Film Fare," *The Silent Worker*, July 1962.
30. "Film Fare," 1962.
31. Gough, *Remarks Prepared*.
32. Gough, *Remarks Prepared*.
33. Gough, *Remarks Prepared*.
34. Gough, "Captioned Films for the Deaf," *School Life* (January–February 1962): 26.
35. Gough, *Remarks Prepared*.

36. Charlayne Hunter-Gault, "Remembering Ed Muskie." *Online NewsHour*. March 26, 1996. http://www.pbs.org/newshour/bb/remember/muskie_3-26.html.
37. Bill Stark, "The Logic of the Motion Picture in the Classroom: Films in Schools for the Deaf (1915–1965)," *Described and Captioned Media Program*, https://dcmp.org/learn/254-the-logic-of-the-motion-picture-in-the-classroom-films-in-schools-for-the-deaf-1915–1965.
38. Captioned Films for the Deaf, Hearing Before a Special Subcommittee of the Committee on Labor and Public Welfare (Washington, DC: US Government Printing Office, 1962), 2. https://dcmp.org/learn/static-assets/nadh248.pdf.
39. Prepared statement by Maurice B. Mitchell, president, Encyclopaedia Britannica Films, Inc., Wilmette, Ill., Captioned Films for the Deaf, Hearing Before a Special Subcommittee of the Committee On Labor and Public Welfare, United States Senate, Eighty-Seventh Congress, Second Session on S.2511, A Bill to Provide for an Increased Program of Captioned Films for the Deaf (Washington, DC: US Government Printing Office, 1962), 6. https://dcmp.org/learn/static-assets/nadh248.pdf.
40. Muskie continued, "During the past several years I have been interested in the potential of modern educational techniques. Because of this interest, I have consulted with experts in the field of audiovisual aids and other teaching devices. The more I have studied this field, the more I have been convinced of the tremendous possibilities in revolutionary improvements in educational programs, especially in such specialized fields as the training of the deaf. Bill Stark, "The Logic of the Motion Picture in the Classroom: Films in Schools for the Deaf (1915–1965)," *Described and Captioned Media Program*, https://dcmp.org/learn/254-the-logic-of-the-motion-picture-in-the-classroom-films-in-schools-for-the-deaf-1915–1965.
41. Gough also mentioned uses of television teaching machines, and programmed learning with the deaf. Gough, *Remarks Prepared.*
42. Kundert, "Captioned Films," 83.

## Chapter 7

1. Kundert, "Captioned Films for the Deaf," 97.
2. Malcolm J. Norwood, "Captioned Films for the Deaf," *Exceptional Children* 43, no. 3 (1976). Reproduced by Described and Captioned Media Project. https://dcmp.org/learn/84-captioned-films-for-the-deaf.
3. John A. Gough, "Captioned Films for the Deaf: The New Program," *Volta Review*, January 1963.
4. Doin Hicks, "Captioned Films for the Deaf Curriculum Workshops." *Described and Captioned Media Project.* https://dcmp.org/learn/376-captioned-films-for-the-deaf-curriculum-workshops
5. Gil Delgado was a decorated military veteran. His awards included the WWII Asian-Pacific Campaign Medal, the WWII Victory Medal, and the Good Conduct Medal. Delgado had earned a bachelor of arts degree from St. Michael's College, Santa Fe, a master of science degree from Gallaudet College, and a PhD from Catholic University. He had experience with the summer faculty at Indiana School for the Deaf and had taught at the California School for the Deaf at Riverside and the New Mexico School for the Deaf in Santa Fe, and he served as principal of the high school at the California School for the Deaf at Berkeley. Delgado had also

edited four books on deafness, including some that center on deaf Latinos. Delgado married Cecilia Ortiz in 1949 and they had six children. Cecilia died of cancer in 2001. Symposium on Research and Utilization of Education Media for Teaching the Deaf.

6. "John Gough" (unpublished notes by Gil Delgado), n.d.
7. Alice L. Young, Harding School, in Kundert, "Captioned Films for the Deaf," 94.
8. Some twenty deaf teachers in attendance at a summer mathematics workshop responded favorably to a captioned teacher training film. Produced at the University of Southern California and captioned by the Office of Education, the film is one of a series designed to acquaint teachers with the new methods of teaching mathematics. Leon Auerbach, mathematics professor at Gallaudet College, arranged to make the film available to the workshop.
9. "College Gets $45,945 Grant for Deaf Films," *The Star Press* (Muncie, Indiana), December 17, 1964, 25.
10. "Session for Deaf Set at Ball State," *Muncie Evening Press* (Muncie, Indiana), July 17, 1963, 10. The third summer that a captioned film program was held on the Ball State campus, it was funded by contract from the US Office of Health, Education, and Welfare for Phase 3 of the program. Ball State received $45,945, announced by Senators Birch Bayh (D-IN) and Vance Hartke (D-IN) on December 16, 1964. The grant would finance a six-week program for twenty teachers of deaf students from state and private schools from throughout the country.
11. Boyce Williams, "Vocational Rehabilitation Services for the Deaf," *American Annals of the Deaf* (October 1964).
12. Williams, "Vocational Rehabilitation Services."
13. Participants received a stipend and could opt to stay in a college residence hall. Also, those who wished could take the workshop for three semester hours of credit. The Ball State faculty representative would be responsible for reviewing and approving work and grade reports.
14. Hicks grew up in rural south Missouri on a small dairy and truck farm. He attended local public schools and then Lyon College in Batesville, Arkansas. Following receipt of a liberal arts degree, he received a Ford Foundation fellowship to study the education of deaf students and followed that with graduate work in secondary education at the University of Arkansas. He then received a two-year fellowship from the Southern Education Foundation for graduate study culminating in the receipt of a doctorate in education. Following graduate school was a position as director of the Pilot School for the Deaf in Dallas, Texas. During his tenure there he also directed a major federal grant program for regional services to deafblind children as well as a grant from CFD to individualize instruction through various strategies including creative ways to use visual media. Doin Hicks, "Captioned Films for the Deaf Curriculum Workshops," *Described and Captioned Media Program*, https://dcmp.org/learn/376-captioned-films-for-the-deaf-curriculum-workshops.
15. Hicks held a new position as an administrator at Gallaudet University in Washington, DC, which gave him responsibility for implementing a new federal law, The Model Secondary School for the Deaf Act, PL 89-694, which provided for the design, construction and staffing of the school to be built on the university campus with plans to serve primarily the District of Columbia and the surrounding five-state region. The lessons he learned at Ball State and Dallas guided much of his actions in

this project, which was soon followed by a similar federal law authorizing the redeveloping of the Kendall Elementary School on campus into a national demonstration elementary school serving a local population. The design, construction, and staffing of these two schools utilized all of the known and available relevant principles of spatial design, media utilization, and training of staff in individualized instruction methodology. Prior to his arrival, the Kendall School had been funded by CFD for an experimental project using computer technology. Utilizing old converted teletypes as computer terminals and a direct telephone hookup to the mainframe computer at Stanford University of California, students were provided a modest program that provided practice in learning basic arithmetic computation. This, of course, was several years before desktop computers were available for general use in classrooms. Hicks, "Captioned Films."

CHAPTER 8

1. Gough, "Captioned Films for the Deaf: Report to CEASD."
2. This was carried out under a contract with the Convention of American Instructors of the Deaf.
3. Len Novick, "Captioned Films for the Deaf: My Perspective," *Described and Captioned Media Program*, https://dcmp.org/learn/375-captioned-films-for-the-deaf-my-perspective.
4. The film was produced by the E.C. Brown Trust of Portland, Oregon. "Film Fare," *The Silent Worker*, October 1964, 7.
5. The number of depositories varied between fifty-eight and sixty over the years.
6. Mike Waugh, "Captioning Techniques for the Captioned Films/Videos Captioning Workshop: Part 1," Manual prepared for use in the Captioned Films/Videos Program, US Office of Education, Office of Special Education, *Described and Captioned Media Program*, https://dcmp.org/learn/100-captioning-techniques-for-the-captioned-films-videos-captioning-workshop-part-i.
7. Waugh, "Captioning Techniques."
8. Bill Stark, "Described and Captioned Media Program: Seventy Years of Progress," *Described and Captioned Media Program*, https://dcmp.org/learn/379-described-and-captioned-media-program-seventy-years-of-progress.
9. *The Deaf American*, January, 1965.
10. Kundert, "Captioned Films for the Deaf," 164.
11. Kundert, "Captioned Films for the Deaf," 164–65.
12. Kundert, "Captioned Films for the Deaf," 166.
13. Kundert, "Captioned Films for the Deaf," 137.
14. "Workshop for Deaf Interpreters Held," *The Star Press* (Muncie, Indiana), June 18, 1964, 10.
15. Kundert, "Captioned Films for the Deaf," 111.
16. Kundert, "Captioned Films for the Deaf,"140.
17. Bill Stark, email message to author, July 1, 2018.
18. PL 89-258, http://uscode.house.gov/statutes/pl/89/258.pdf.
19. "MSCF Program Developments FY 1966," Gallaudet University Archives.
20. Available film titles were distributed through the CFD catalog. In 1967, for example, the catalog "The Captioned Films for the Deaf Program" was coordinated

by James J. Kundert, a deaf assistant education specialist, Captioned Films for the Deaf.

21. John A. Gough, "Gullible's Travels" (unpublished personal memoirs of John A. Gough), shared by Susan Gough Updegraff, January 2019, 93.
22. Donald G. Ferguson, "A Little Bit About Captioned Films for the Deaf," *Rocky Mountain Leader*, December 1967.
23. Stark, "Described and Captioned Media Program: Seventy Years."
24. William T. Gladmon from WQED-WQEX was appointed director, and WPSD superintendent Sam B. Craig was codirector.
25. March 17–19. William D. Jackson, "Southern Regional Media Center for the Deaf." *American Annals of the Deaf* 114 (November 1969): 841–44.
26. Within a few years, this same wishful thinking would be offered when officers were chosen for the new organization, Telecommunications for the Deaf, Inc. (now known as Telecommunications for the Deaf and Hard of Hearing, Inc., or TDI) to lead the implementation of a teletypewriter/modem network that would make the telephone accessible to deaf people. See Lang, *A Phone of Our*.
27. Malcolm J. Norwood, "Captioned Films for the Deaf," *Exceptional Children* 43, no. 3 (1976).
28. "Workshop Ends on Use of Films in Deaf Schools" *Arizona Daily Star*, November 10, 1965, 4. Concurrently with the Ball State workshops, in June 1964, Gough had participated in a pioneering Workshop on Interpreting for the Deaf at Ball State. It was financed by the Vocational Rehabilitation Administration and directed by William J. McClure, superintendent of the Indiana School for the Deaf, and helped to set standards for the young profession. Seventy participants from twenty-six states attended, and Gough had played a role in setting it up. Training films for interpreters was one of the topics.
29. Lucille Hutchinson, "U.S. Education Office to Use NU for Production of Deaf Materials," *The Lincoln Star* (Lincoln, Nebraska), October 27, 1965, 1.
30. "National Captioning Institute: Commemorative History," http://www.cybrations.com/Clips/NCI_Chapter_1.htm.
31. *Florida School Herald*, newsletter clipping of the Florida School for the Deaf and Blind, December 1964. Gallaudet University Archives.
32. George Propp, *Missouri Record*, December 1967, The Midwest Regional Media Center for the Deaf Administrative Assistant, Regional Coordinator.
33. George Propp, "An Overview of Progress in Utilization of Educational Technology for Educating the Hearing Impaired," 1978 Symposium on Research and Utilization of Educational Media for Teaching the Deaf, Captioned Films and Telecommunications Branch and Media Development Project for the Hearing Impaired, University of Nebraska, Lincoln.
34. "Third Symposium on Research and Utilization of Educational Media for Teaching the Deaf," Lincoln, Nebraska, April 10–12, 1967.
35. In acquisition, equipping classes with media was not all done by Captioned Films. PL 89-313 supplied more funding to state-supported schools.
36. John A. Gough, "Report from Captioned Films for the Deaf" (paper prepared for the Symposium on Research and Utilization of Educational Media for Teaching the Deaf: The Educational Media Complex, Nebraska Center for Continuing Education,

Lincoln, Nebraska, February 5–7, 1968), *Described and Captioned Media Program*, https://dcmp.org/learn/static-assets/nadh92.pdf.

37. Gough, "Report from Captioned Films."
38. The 1969 Symposium on Educational Media for the Deaf, for example, included such topics as mediated self-instruction, computer-assisted instruction, and use of programmed instruction. It was held in the Graham H. Anthony Vocational Rehabilitation Center in Hartford, CT.
39. Art Kruger, "A Day to Remember for Leon, LACD," *The Deaf American*, 21, no. 9 (May 1969), 37.
40. In recognition of his contributions, Ball State University, his alma mater, conferred upon John Gough the honorary degree of Doctor of Laws in 1966.
41. The entertainment collection was housed in a warehouse-type facility in downtown Indianapolis. It was a labor-intensive, two-shift operation, with many of the employees being graduates of the Indiana School for the Deaf. The operation was overseen by Val Brummett Jr.
42. In March 1970, a year after his own retirement, he joined 300 educators, staff members, and students at the American School for the Deaf to honor Edmund B. Boatner and Maxine T. Boatner for their dedicated work since the mid-1930s. Edmund had served as the first president of Captioned Films for the Deaf, Inc., which had become the agency directed by Gough in the US Office of Education.
43. "Film Fare," *The Deaf American*, July–August 1970.
44. "Statement of Hon. Edmund S. Muskie, U.S. Senator from the State of Maine," Captioned Films for the Deaf. Hearing Before a Special Subcommittee of the Committee on Labor and Public Welfare, United States Senate, Eighty-Seventh Congress, Second Session on S. 2511, A Bill to Provide for an Increased Program of Captioned Films for the Deaf, August 7, 1962, *Described and Captioned Media Program*, https://dcmp.org/learn/static-assets/nadh248.pdf.
45. After years of Gough's staff attempting to set up a new program, one finally began in 1970 at Syracuse University in upstate New York. The program started with five deaf and hearing Fellows, and five additional Fellows were added in 1971. It was funded for seven years and the last participant graduated in 1981. By then, twenty-two participants had completed a graduate sequence of courses and half had received doctoral degrees. Delgado personally recruited many of these participants, including Bob Davila, a deaf Hispanic man who went on to become the first individual with a disability to serve as assistant secretary in the Office of Special Education and Rehabilitative Services (US Department of Education), as well as vice president at the National Technical Institute for the Deaf at Rochester Institute of Technology, and president of Gallaudet University. Another deaf Hispanic graduate, Ramon Rodriguez, became a US Department of Education program specialist. Among other graduates of the Educational Technology doctoral program were Barbara Bodner-Johnson and Elaine Costello, who were sent to Gallaudet University in Washington, DC, and Jim DeCaro, Paul Peterson, and Jim Cox, who were sent to the National Technical Institute for the Deaf at the Rochester Institute of Technology. Also see Harry G. Lang, Oscar Cohen, and Joseph Fischgrund, *Moments of Truth: The Journey of Robert R. Davila, Deaf Educator* (Rochester, NY: Rochester Institute of Technology Cary Press, 2007).

46. The panel Delgado introduced included Raymond Wyman, William Jackson, Marshal Hester, George Propp, Robert Root, and Glen Pfau.
47. *Proceedings of the Symposium on Research and Utilization of Educational Media for Teaching the Deaf* (Lincoln, Nebraska, March 16–18, Sponsors: University of Nebraska, Midwest Regional Center for the Deaf and Southern Regional Media Center for the Deaf with support from the US Department of Health, Education, and Welfare, 1970). https://eric.ed.gov/?id=ED043188.
48. Symposium on Research and Utilization of Educational Media for Teaching the Deaf, Lincoln, Nebraska, March 17–19, 1969. Sponsors: University of Nebraska and Midwest Regional Center for the Deaf with support from the US Department of Health, Education, and Welfare, 1969). https://eric.ed.gov/?id=ED043987.
49. Edward L. Palmer, "Television's Neglected Strengths." *American Annals of the Deaf*, 115, 1970, pp. 601–4. www.jstor.org/stable/44393423.
50. William Jackson, "1970 Survey of Instructional Television in Programs for the Deaf," *Proceedings of Symposium on Research and Utilization of Educational Media for Teaching the Deaf* (6th symposium, Lincoln, Nebraska, March 16–18, 1970), 171–80.

## Chapter 9

1. *Great Falls Tribune* (Montana), July 10, 1969.
2. Mac also faced a temporary staff shortage a month after he began in his new role. While walking in the early morning hours, his deaf assistant James J. Kundert was attacked from behind and hospitalized. Fortunately, Kundert recovered and returned to his responsibilities in Media Services and Captioned Films (MSCF).
3. Malcolm J. Norwood, "Keynote Address," First National Conference on Television for the Hearing-Impaired. Southern Regional Media Center for the Deaf at the University of Tennessee, Knoxville, December 14–16, 1971.
4. Norwood, "Keynote address."
5. "Film Fare," *The Silent Worker*, January 1963.
6. "Film Fare," 1963.
7. See Lang, *A Phone of Our Own*.
8. Malcolm J. Norwood, "The Second Decade," *American Annals of the Deaf* 114, no. 5 (1968): 824–828.
9. William Jackson, "1970 Survey of Instructional Television in Programs for the Deaf," *Communicative Television for the Deaf Student*, Proceedings of Symposium on Research and Utilization of Educational Media for Teaching the Deaf, ed. George Propp (6th symposium, Lincoln, Nebraska, March 16–18, 1970).
10. Malcolm J. Norwood, "The Involvement of Deaf Persons in Services to the Deaf," *Proceedings of International Congress on Education of the Deaf* (Stockholm: Sveriges lärarförbund, 1972).
11. Norwood, "The Involvement of Deaf Persons."
12. Malcolm J. Norwood, "Media Services and Captioned Films." *Described and Captioned Media Program*, https://dcmp.org/learn/141-media-services-and-captioned-films
13. Research emphases were expanding across the nation at a remarkable rate, as had

become evident at the sixth Nebraska symposium held in Lincoln in March 1970. Norwood's responsibilities were also expanding in the postsecondary arena. In addition to responsibility for the PhD program at Syracuse, he was named, in February 1971, by the US Department of Health, Education, and Welfare (HEW) the liaison officer for the National Technical Institute for the Deaf (NTID) at Rochester Institute of Technology. Edwin W. Martin, associate commissioner of the Bureau of Education for the Handicapped in HEW, explained that Norwood would continue to serve as assistant chief of Media Services and Captioned Films in the US Office of Education. "Norwood . . . .has made an outstanding contribution to education of the deaf, and has been a leader in securing recognition that deaf citizens can play key roles in American government and business." As liaison to NTID, he would be able to tap the growing research potential at this new college.

14. In May 1972 Norwood's accomplishments were recognized by Gallaudet College, which granted him an honorary Doctor of Laws degree.
15. Karen Brickett, "An Interview with Dr. J. Norwood," December 1979, *Described and Captioned Media Program,* https://dcmp.org/images/learning_center/535/mac-norwood-lesson-guide.pdf.
16. Brickett, "An Interview."
17. Eugene W. Peterson, "The Editor's Page: Expansion of Media Services-Captioned Films," *The Deaf American,* 23, February 1971.
18. When the National Association of Broadcasters set up a special subcommittee for captioning television, it was concerned mainly with the development of the engineering specification standards necessary to set up a television captioning system for broadcasting purposes.
19. "Demonstration Shows Use of Captioned Television Broadcast," *The Buff and Blue,* 81, no. 9, February 17, 1972. In analog television systems in the 1970s, the "vertical blanking interval," an interval of time between the last line of a frame and the beginning of the next frame, was used to carry digital data. Nothing sent during the interval was displayed on the screen. In US analog broadcast television, this required the last forty-five lines of each 525-line frame. "Line 21" was reserved for captioning data.
20. Marc Okrand, "Captioned Television," *Gallaudet Encyclopedia of Deaf People and Deafness, Vol. 3,* ed. J. V. Van Cleve (New York: McGraw-Hill, 1987), 265.
21. Due to the illegibility of fonts in the original two decoder generations, offline closed captions were, in the early days, in all uppercase letters, and some deaf viewers believed this slowed down the reading rate. Joe Clark, "Alphabetized Comments on CAB Captioning Manual," Media Access, https://joeclark.org/access/captioning/cab/cab-all-alpha.html.
22. Okrand, "Captioned Television."
23. Strauss, *A New Civil Right,* 209.
24. "Demonstration Shows Use of Captioned," 1972.
25. Malcolm J. Norwood, "Captioning for Deaf People: An Historical Overview," *Described and Captioned Media Program,* https://dcmp.org/learn/80-captioning-for-deaf-people-an-historical-overview.
26. Strauss, *A New Civil Right,* 207.

27. "Programming Designed for Deaf Television Viewers," *The Des Moines Register* (Des Moines, Iowa), July 23, 1972, 117.
28. *The Indianapolis Star* (Indianapolis, Indiana), June 27, 1973.
29. "Nanette Fabray Speaks Up for Deaf," *The Indianapolis News*, June 26, 1973, 8.
30. Okrand, "Captioned Television," 264–67.

## Chapter 10

1. Excerpts from the *Educational Captioned Films Depository Manager's Handbook* (Indiana University Audio-Visual Center, Bloomington, Indiana, 1977).
2. Malcolm J. Norwood, "Captioned Films for the Deaf," *Exceptional Children* 43 (1976).
3. "Public Law 94-142, The Education for All Handicapped Children Act: Its Development, Implementation, and Current Issues," http://digitalcollections.library.cmu.edu/awweb/awarchive?type=file&item=673247.
4. "Congress Recognizes 10th Anniversary of Public Law 94-142," *Congressional Record,* Wednesday, September 25, 1985; 131, no. 122, https://nfb.org/images/nfb/publications/fr/fr5/issue1/f050112.html.
5. "Congress Recognizes 10th Anniversary," *Congressional Record.*
6. In 1976, the MSCF program developed regulations for loan service, according to the plans outlined in Public Laws 87-715 (Loan Service of Captioned Films for the Deaf) and 91-230 (Elementary and Secondary Education Amendments of 1969).
7. Malcolm J. Norwood, "Future Trends," *American Annals of the Deaf* 27 (1974): 619–623.
8. "Audio Visual Communication and the Education of the Deaf in the United States," Proceedings of International Congress on Education of the Deaf, Stockholm, Sweden, 1970. "Media, Its Role in the Education of the Deaf in the United States" (paper presented at World Federation of the Deaf, Paris, France, 1971). "We Have a Job to Do," Keynote Address, National Conference on Television for the Hearing Impaired, Knoxville, Tennessee, 1971. "New Horizons," *Improving Human Performance: A Research Quarterly* 1, no. 3 (1972): 5–7. "Present and Future Developments" (paper presented at the Conference of Executives of American Schools for the Deaf, Toronto, 1972). "The World of Deafness—As Viewed by an Insider," (paper presented at the annual meeting of the Acoustical Society of America, Buffalo, New York, 1972). "Total Opportunity" (paper presented to meeting of State Teachers of Hearing Impaired, Hartford, Connecticut, 1972). "Media Horizons" (paper presented at Convention of American Instructors for the Deaf, Indianapolis, 1973). "Are We Losing the Media Game?" (paper presented at meeting of Conference of Executives of American Schools for the Deaf, Tucson, Arizona, 1974). "A Look Into the Future of Technology and the Education of the Handicapped" (paper presented at the Convention of the Association for Educational Communications and Technology with Paul Andereck, Atlantic City, NJ, 1974). "Future Trends," *American Annals of the Deaf* 119, no. 5 (1974): 619–623. "Instructional Technology: Its Emerging Role in the Education of the Handicapped" (paper presented at the Symposium of Applied Learning Systems Curriculum Requirements, American University, Washington, DC, 1975). "The Government as a Change Agent: The Federal Perspective"

(paper presented at the Convention of American Instructors for the Deaf, Greensboro, North Carolina, 1975). "Television, Instructional Media and Learning" (paper presented at the VII World Federation of the Deaf, Washington, DC, August 1975). "Media and Technology: It's Impact Educationally and Socially on the Deaf in the United States" (paper presented at the International Congress on the Education of the Deaf, Tokyo, Japan, 1975).

9. David Strom, email to author, December 2, 2018.
10. Bernard Hurwitz, email to author, January 5, 2019.
11. Jerry Nichols, email to author, December 26, 2018.
12. Jeff Hutchins, email to author, November 6, 2019.
13. Hutchins, email to author.
14. Hutchins.
15. Malcolm J. Norwood, "Captioning for Deaf People: An Historical Overview," *Described and Captioned Media Program*, https://dcmp.org/learn/80-captioning-for-deaf-people-an-historical-overview. Excerpted from *Speech to Text: Today and Tomorrow*, ed. J. E. Harkins and B. M. Virvan, (Washington, DC: Gallaudet Research Institute, 1988).
16. Gail L. Kovalik, "'Silent' Films Revisited: Captioned Films for the Deaf," *Library Trends* 41 (1992): 100–117.
17. Malcolm J. Norwood, "Media Services and Captioned Films," *Described and Captioned Media Program*, https://dcmp.org/learn/141-media-services-and-captioned-films.
18. Carl Jensema, email to author, November 4, 2019.
19. Until 1981, the insertion equipment (called "encoders") worked only with prerecorded programs. All data sent to an encoder was placed on line 21, two bytes per frame. (Video then was thirty frames per second, which meant the maximum data rate was sixty bytes per second.) With live feeds, it was necessary to invent a "smart encoder" that could pass through existing line 21 data (such as captions for a commercial) while also inserting new data provided in real time. And they also had to invent a way to create captions that were a full and accurate transcript of the audio portion of the show. Hutchins was responsible for developing these systems for live captioning. Jeff Hutchins, email to author, November 6, 2019.
20. Stephen Good, "Technology Brings TV to the Deaf," *The Courier-Journal* (Louisville, Kentucky), May 18, 1980, 67.
21. Norwood, "Captioning for Deaf People."
22. Norwood, "Captioning for Deaf People."
23. Strauss, *A New Civil Right.*
24. Len Novick, email to author, November 4, 2019.
25. Howard M. Quigley, the first executive director of CEASD, was simultaneously project director of the Captioned Films for the Deaf Distribution Center from 1966 to 1975. Len Novick, email to author, November 28, 2019. A good reference for CEASD's early involvement in CFD, a cogent summary of CFD's early history, including Howard M. Quigley's role, and other matters can be found here: Richard G. Brill, *The Conference of Educational Administrators: A History* (Washington, DC: Gallaudet College Press, 1986). Specifically, see chapters 2 and 3.
26. Leonard Novick, "Microcomputer Activities of the Special Materials Project,"

*Described and Captioned Media Project*, https://dcmp.org/learn/static-assets/nadh202.pdf.

27. Novick, "Microcomputer Activities."
28. Peggy Thomson, "Today's Education. Closed Captions for TV: Opening a New World to the Hearing Impaired," *Journal of the National Education Association* 70, no. 3 (September–October 1981): 52–54. https://digitalcollections.library.cmu.edu/portal/awarchive?type=file&item=578670.
29. Thomson, "Closed Captions."
30. The budget for captioning was increased to a high of $424,000 in 1985.
31. Unfortunately, time and budgets prohibited them from taking that approach beyond the experimental level.

## Chapter 11

1. Richard C. Levy, "TV Opens Up Line to the Deaf," *Parade*, July 8, 1979, 10.
2. Levy, "TV Opens Up," 10.
3. Bill Stark, "How Bird Hunting in North Carolina Saved Captioning," *Described and Captioned Media Program*, https://dcmp.org/learn/36-how-bird-hunting-in-north-carolina-saved-captioning.
4. Lee Margulies, "TV Programming for Deaf on Upswing," *The Journal Herald* (Dayton, Ohio), December 21, 1983, 39.
5. Lee Margulies, "Technology Enables Deaf to Enjoy More on Television," *The Morning News* (Wilmington, Delaware), December 11, 1983, 71.
6. Strauss, *A New Civil Right*, 219.
7. John Koskinen, email to author, November 6, 2019.
8. Koskinen.
9. Levy, "TV Opens Up," 15.
10. Levy, "TV Opens Up," 15.
11. Peggy Thomson, "Closed Captions," 52–54, https://digitalcollections.library.cmu.edu/portal/awarchive?type=file&item=578670.
12. Alan Hurwitz, email to author, September 25, 2018.
13. Thomson, "Closed Captions."
14. Malcolm J. Norwood, "Captioning for Deaf People: An Historical Overview," *Described and Captioned Media Program*, https://dcmp.org/learn/80-captioning-for-deaf-people-an-historical-overview.
15. Thomson, "Closed Captions," 52–54,
16. Thomson, "Closed Captions," 52–54.
17. Thomson, "Closed Captions," 52–54.
18. Thomson, "Closed Captions."
19. Levy, "TV Opens Line," 15.
20. Levy, "TV Opens Up," 15.
21. Norwood, "Captioning for Deaf People."
22. Strauss, *A New Civil Right*, 211.
23. As described by Strauss in *A New Civil Right*, 212.
24. "Deaf Protest CBS's Refusal to Use Captions," *The Daily Spectrum* (Saint George, Utah), May 20, 1982, 9.
25. "Deaf Demonstrate Against CBS," *The Billings Gazette* (Billings, Montana), May

20, 1982, 6.

26. John Carlson, "Iowans Protest CBS Affiliates' Caption Policy," *The Des Moines Register* (Des Moines, Iowa), May 20, 1982, 3.
27. Norman Black, "NBC Drops Captions for Deaf," *The News-Journal Papers* (Wilmington, Delaware), August 26, 1982, D6.
28. Strauss, *A New Civil Right*, 218.
29. *The Jackson Sun* (Jackson, Tennessee), November 7, 1987, 2.
30. Gregory J. Downey, *Closed Captioning: Subtitling, Stenography, and the Digital Convergence of Text with Television* (Baltimore: Johns Hopkins University Press, 2008), 51–52.
31. Adam R. Nelson and John L. Rudolph (eds), *Education and the Culture of Print in Modern America* (Madison: University of Wisconsin Press, 2010), 203.
32. John Koskinen, email to the author, November 6, 2019.
33. Norwood, "Captioning for Deaf People."
34. National Captioning Institute, "History of Closed Captioning," https://www.ncicap.org/history-of-cc.
35. Charles R. Babcock, "Deaf Attorney Argues Before High Court for Disabled Student," *Washington Post*, March 24, 1982, https://www.washingtonpost.com/archive/politics/1982/03/24/deaf-attorney-argues-before-high-court-for-disabled-student/3effd1c7-1b0b-4303-b6fb-e3c352d83c96/?utm_term=.4e981154c761.
36. In one study conducted by the author's graduate student in 2004, five hours of live captioned television were examined, including thirty-minute segments from ten different live news programs randomly selected. Each error was recorded as it was detected and a list of errors, along with the corresponding spoken messages were compiled. A total of 235 caption errors were identified with an average of 23.5 errors on each station. The most common error categories were spelling errors (33 percent), word substitution errors (19 percent), and phonetic system errors (13 percent). In drawing implications for teachers, Melissa VanZant wrote, "It is critical that deaf viewers have the skills necessary to mentally repair these errors, when possible, in order to comprehend live news programs. Enhanced comprehension through repair strategies may enhance incidental learning" (Melissa VanZant, "An Analysis of Real-Time Captioning Errors" [Master's thesis, Rochester Institute of Technology, 2005]). See also Martin H. Block and Marc Okrand, "Real-Time Closed-Captioned Television as an Educational Tool," Symposium on Research and Utilization of Educational Media for Teaching the Deaf, *American Annals of the Deaf* 128, no. 5 (1983), *Described and Captioned Media Program*, https://dcmp.org/learn/static-assets/nadh149.pdf.
37. "'Silent' Films Revisited; Captioned Films for the Deaf," *The Free Library*, https://www.thefreelibrary.com/%22Silent%22+films+revisited%3a+captioned+films+-for+the+deaf.-a013294380.
38. "Captioning and CART," Hearing Loss Association of America, https://www.hearingloss.org/hearing-help/technology/cartcaptioning/.
39. Strauss, *A New Civil Right*, 218.
40. E. Ross Stuckless, "Real-Time Graphic Display and Language Development for the Hearing-Impaired," *Volta Review* 83, no. 5 (1981): 291–300; E. Ross Stuckless, "Real-Time Transliteration of Speech into Print for Hearing-Impaired Students in Regular Classes," *American Annals of the Deaf* 128, no. 5 (1983):

619–24.

41. See, for example, M. S. Stinson, "The Importance of Technologies in the Education of Deaf Students," in *Evidence-Based Practices in Deaf Education*, ed. M. Marschark and H. Knoors (New York: Oxford University Press, 2018), 543–68; and M. S. Stinson, "Current and Future Technologies in the Education of Deaf Students," in *Oxford Handbook of Deaf Studies, Language, and Education*, ed. M. Marschark and P. Spencer (New York: Oxford University Press, 2010), 93–110.
42. "Caption First: The Gold Standard in Communication Access," http://captionfirst.com/services/cart/.
43. Norman Black, "ABC Captions Morning News Shows," *Tallahassee Democrat* (Tallahassee, Florida), November 4, 1984, 97.
44. "U.S. to Increase Aid for Captioning TV for the Deaf," *Star Tribune* (Minneapolis, Minnesota), July 5, 1984, 35.
45. "NCI Recap'd," National Captioning Institute, http://www.ncicap.org/services/recapd/.
46. Lisa Twyman, "At Last, the Deaf Can Follow All of the Monday Night Game Commentary," *Sports Illustrated*, December 23, 1985, https://www.si.com/vault/1985/12/23/622666/at-last-the-deaf-can-follow-all-of-the-monday-night-game-commentary.
47. "Rodney Wilson Walker, *My Life Story*, 2006," as cited by Jodi B. Kinner and edited by Valerie G. Kinney, in *United Utah Organizations of the Deaf*, 2012, https://www.utahdeafhistory.com/uploads/5/4/2/6/5426987/united_utah_organizations_deaf.pdf.
48. McCay Vernon, "Dr. Malcolm Norwood Retires," *American Annals of the Deaf* 132 (December 1987): 361.
49. Salvatore J. Parlato, Jr., "Nonverbal Films: Guidelines for Their Utilization With Deaf Learners," Symposium on Research and Utilization of Education Media for Teaching the Deaf, *Described and Captioned Media Program*, https://dcmp.org/learn/static-assets/nadh118.pdf.
50. Richard Hack, "Televisions," *Hollywood Reporter*, April 15, 1988.
51. Hack, "Televisions."
52. Hack, "Televisions."
53. Hack, "Televisions."
54. Len Novick, "Captioned Films for the Deaf: My Perspective," *Described and Captioned Media Program*, https://dcmp.org/learn/375-captioned-films-for-the-deaf-my-perspective.
55. Novick, "Captioned Films."
56. Strauss, 230. See Dennis Kelly, "TV Closed-Captions Fight Illiteracy," *USA Today*, July 11, 1990, 6D.
57. Patricia S. Koskinen, Robert M. Wilson, Linda B. Gambrell, and Carl Jensema, *Using the Technology of Closed-Captioned Television to Teach Reading to Handicapped Students. Performance Report* (Falls Church, VA: NCI, February 1987), https://eric.ed.gov/?id=ED314891.
58. Strauss, *A New Civil Right*, 220–21.
59. Novick, "Captioned Films."
60. Strauss, *A New Civil Right*, 229.
61. Frank Bowe, "Despite Gallaudet Protest Education for the Deaf Remains Inadequate

and Discriminatory," *The News Journal* (Wilmington, Delaware), March 26, 1989, 67.

62. Strauss, *A New Civil Right*, 229.
63. Karen Peltz Strauss provides much detail to the decoder chip bill in chapter 10 of *A New Civil Right*. She writes that there were fourteen cosponsors to the bill.
64. See Strauss, *A New Civil Right*, 230, 242n17.
65. Strauss, *A New Civil Right*, 32.
66. Karen Peltz Strauss, email to author, February 23, 2020.
67. Strauss provides much more detail in chap. 10 of *A New Civil Right*.
68. Strauss provides much more detail in chap. 10 of *A New Civil Right*.
69. Strauss, *A New Civil Right*, 236.
70. "New Law Will Expand TV Captions for the Deaf," *New York Times*, October 16, 1990, 22.
71. Hack, "Televisions."

## Chapter 12

1. Ernest E. Hairston and Linwood Smith, *Black and Deaf in America: Are We That Different?* (Silver Spring: TJ Publishers, 1983); Lang and Meath-Lang, *Deaf Persons*.
2. "History of Closed Captioning," National Captioning Institute, https://www.ncicap.org/history-of-cc.
3. Nicole Snell, "Toward a Theory of Media Reconciliation: An Exploratory Study of Closed Captioning" (PhD diss., Clemson University, 2012), *All Dissertations*, 975, https://tigerprints.clemson.edu/all_dissertations/975.
4. MTPS relocated the movie and special interest depository (library) to Shawnee Mission, Kansas.
5. Salvatore Parlato Jr., "Re-discovering Films with Captions," *Teaching English to Deaf and Second Language Students* 3, no. 1 (1985): 17–20.
6. For more details on the life and work of Robert R. Davila, see Harry G. Lang, Oscar Cohen, and Joseph Fischgrund, *Moments of Truth: The Journey of Robert R. Davila, Deaf Educator* (Rochester, NY: Rochester Institute of Technology Cary Graphic Arts Press, 2007).
7. Stark began working with CFD in the 1960s as a film depository manager at the Illinois School for the Deaf. His formal education began in a one-room school house in a rural section of southern Illinois known as Little Egypt. From a first-grade class with an enrollment of two to a consolidated high school senior class of fifty, Bill was afforded ample opportunities and encouragement to develop a love of books and films. An incredibly caring and skilled school librarian was responsible for the former, and it was an English Literature teacher who showed a black-and-white film of *Romeo and Juliet*, thereby helping Bill and his classmates to understand both the power of film and Shakespearian language for the first time. As the son of a sharecropper and from a family of virtually no college graduates, Bill dreamed of one day attending nearby Southern Illinois University (SIU) and nurturing his love of books and films. A teaching scholarship made possible earning a bachelor of arts degree there, just as a graduate assistantship later provided opportunity at SIU for advanced degree work. For both degrees, Bill's concentration was on library science and educational media while earning teaching and administrative certificates.

   In 1966, during his final year at SIU, he was personally recruited by Dr. Frank

Withrow, a researcher of national renown in the use of educational media in special education. Withrow, who became a mentor to Bill, wanted him to head a new multi-million-dollar media project at the Illinois School for the Deaf. Its purpose was to enhance the academic achievement and communication skills of deaf students (K–12). A bonus of accepting the position was the opportunity to offer through his own instruction teacher preparation courses in the deaf education department at nearby MacMurray College. Receiving national attention, his accomplishments over the next two decades included developing a model combined library and audiovisual services program for grades K–12, designing of a facility to house media and utilize these services (reported to be the best of its kind nationwide), and creating innovative approaches in teaching reading and language to reluctant readers. One example of such was the use of comic books in the library and classrooms, this later leading to his development for national distribution of one of the first comic books to feature a deaf superhero named "Blue Ear."

8. Bill Stark, email to author, July 12, 2018.
9. Doin Hicks, Captioned Films for the Deaf Curriculum Workshops, *Described and Captioned Media Program,* https://dcmp.org/learn/376-captioned-films-for-the-deaf-curriculum-workshops.
10. Malcolm J. Norwood, "Just Don't Scramble the Wrong Egg," in *Captioning: Shared Perspectives,* ed. B. Braverman and B. J. Cronin (Rochester, NY: National Technical Institute for the Deaf, 1980), https://dcmp.org/learn/static-assets/nadh83.pdf.
11. Toby R. Silver, email to author, January 29, 2020.
12. "The Harvey Goodstein Collection, 1982–1998," Gallaudet University Archives, https://www.gallaudet.edu/archives-and-deaf-collections/colleuctions/manuscripts/mss-206.
13. Testimony by Toby R. Silver, Chair, Television For All Coalition in Support of H.B. 381 Video Caption Access Bill, February 3, 1994.
14. Silver, "Testimony."
15. Willis Mann, "Legislative Heat . . . and Cold Shoulders," *MDAD News* (March 1994).
16. Toby R. Silver, email to author, January 19, 2020.
17. Gough, "Gullible's Travels."
18. Gough, "Captioned Films for the Deaf," 20.
19. Frank B. Withrow, letter to John A. Gough, June 11, 1992, shared with author by Susan Gough Updegraff, January 2019.
20. Bill Stark, "Described and Captioned Media Program: Seventy Years of Progress (1946–2016)," *Described and Captioned Media Program,* https://dcmp.org/learn/379-described-and-captioned-media-program-seventy-years-of-progress.
21. Stark, "Described and Captioned Media Program."
22. Stark, "Described and Captioned Media Program."
23. Strauss, *A New Civil Right,* 263.
24. "Closed Captioning Begins," TDI Online, http://www.tdi-online.org/tdi/Captioning-History.asp.

### Chapter 13

1. "Films Look to Captioned Audience," https://www.wired.com/2000/07/films-look-to-captioned-audience/.
2. The Society of Motion Picture & Television Engineers formed a working group for

captioning standards on MPEG digital video, which is used on DVDs.

3. James Hyder, "Report on a Test of Closed-Caption Systems in the Langley Theater of the National Air and Space Museum, March 28–April 1, 1994," Washington, DC: National Air and Space Museum, Smithsonian Institution, October 3, 1994.
4. "Closed Captioning at LSC Movies," MIT Lecture Series Committee, http://lsc.mit.edu/info/captions-release.shtml.
5. Kendra Mayfield, "Films Look to Captioned Audience," *Wired*, July 28, 2000, https://www.wired.com/2000/07/films-look-to-captioned-audience/.
6. Roy Miller, "Good, But Not Perfect." *TDI World*, 43, https://tdiforaccess.org/wp-content/uploads/2020/01/Vol.-43-Issue-1-2-2012-Going-to-the-Movies-with-Captions.pdf
7. Miller, "Good," 43.
8. Katherine Bouton, "Gobsmacked by Rudeness. Frustrated by Technology, https://katherinebouton.com/2016/01/21/gobsmacked-by-rudeness-frustrated-by-technology/.
9. "Barbara Montan and Chris Montan," Rochester Institute of Technology, https://www.rit.edu/ntid/radscc/tripod/mediafront-player-youtube/Barbara-Montan-and-Chris-Montan.
10. "TRIPOD—Brief History," Rochester Institute of Technology, https://www.rit.edu/ntid/radscc/tripod/about-tripod/tripod-brief-history.
11. "TRIPOD."
12. Nanci Linke-Ellis, "Accessibility for Digital Cinema—A 2012 Update," *SMPTE Motion Imagining Journal* (September 2012).
13. Linke-Ellis, "Accessibility," 90.
14. Linke-Ellis, "Accessibility."
15. Karen Peltz Strauss, email to author, August 30, 2019.
16. See "Self-Help for Hard of Hearing People," Movie Caption Legal Brief, United States District Court, District of Columbia, https://joeclark.org/dossiers/BallvAMC/CMCcommentsFinal.pdf.
17. "Captioned Movie Access Advocacy—Timeline," *Fookem and Bug* (blog), June 29, 2007, https://fookembug.wordpress.com/2007/06/29/captioned-movie-access-advocacy---timeline/.
18. "First Open Captioned Movie Test Underway!!" July 3, 1997, http://members.tripod.com/maf_/old/captiond.htm.
19. Associated Press, "Hearing-Impaired Seek Encores of 'Open Captioning,'" *Deseret News*, July 21, 1997, https://www.deseret.com/1997/7/21/19324698/hearing-impaired-seek-encores-of-open-captioning.
20. Karen Peltz Strauss, email to author, September 12, 2019.
21. Mark Barna, "Silence in the House: Tripod Captioned Films Sets the Standard for Open Captioning," *In Focus* (April 2001), 16.
22. Mayfield, "Films Look to Captioned Audience."
23. Mayfield, "Films Look to Captioned Audience."
24. Mayfield, "Films Look to Captioned Audience."
25. Linda S. Kadaba, "Advocate Helps Open Arts to the Deaf," *The Daily Oklahoman* (Oklahoma City, Oklahoma), February 4, 2002, 13.
26. Linda K. Harris, "Deaf People Challenge Hollywood," *Philadelphia Inquirer*

(Philadelphia, Pennsylvania), June 4, 2000, 28

27. "Organization Opens the Arts to the Deaf," *The Tribune* (Seymour, Indiana), December 26, 2001, 14.
28. "Cultural Events Available for the Deaf," *The Star Press* (Muncie, Indiana), December 29 2001, 30.
29. "Cultural Events Available," 30.
30. "Organization Opens the Arts," 14.
31. For her advocacy efforts, Finkle and Creative Access received the 1997 James C. Marsters Promotion Award from Telecommunications for the Deaf, Inc., in recognition of outstanding contributions in improving visual accessibility at local movie theaters for deaf and hard of hearing citizens. In 2001 she received the Alan R. Barwiolek Memorial Award and the Connecticut Coalition for Captioning & the Connecticut Commission for the Deaf and Hearing Impaired Award of Appreciation; and in 2004 the Robert M. Greenmum Local Leadership Award from the NAD.
32. Barbara Raimondo, email to author, January 25, 2020.
33. Linda K. Harris, "Deaf People Challenge Hollywood to Caption Pictures," *Philadelphia Inquirer* (Philadelphia, Pennsylvania), June 4, 2000, 4.
34. Bouton, "Gobsmacked by Rudeness."
35. James Jones, email to author, August 24, 2018.
36. The charter members included the Alexander Graham Bell Association for the Deaf and Hard of Hearing, the American Society for Deaf Children, the Association of Late-Deafened Adults, the Cochlear Implant Association, Inc., the Deaf and Hard of Hearing Consumer Advocacy Network, Deaf Seniors of America, the League for the Hard of Hearing, the NAD, Self Help for Hard of Hearing People, Inc., and Telecommunications for the Deaf, Inc.
37. The CMC encouraged its coalition representatives to be actively involved in developing all captioning standards. See TDI's *A Path toward an Accessible World: 1968–2018.*
38. *The GA-SK*, published by TDI, Spring 2000.
39. Jamie Berke, email to author, January 14, 2019.
40. "Tripod Captioned Films Transitions to: InSight Cinema," *Audiology Online*, December 20, 2002, https://www.audiologyonline.com/releases/tripod-captioned-films-transitions-to-5956.
41. Paul Singer, "Hearing-Impaired Get Movie Treat," *Chicago Tribune* (Chicago, Illinois), May 7, 2004, 6.

### Chapter 14

1. Cindy Malone, letter to Jason Stark, December 6, 2017.
2. Casey Zang, letter to Jason Stark, February 28, 2018.
3. Teresa Ramiriz, letter to Jason Stark, October 18, 2017.
4. Jack F. Levesque, "Deaf People Are Patient People." https://dcmp.org/learn/59-deaf-people-are-patient-people.
5. Levesque, "Deaf People."
6. Laura Pulfer, "Tax Dollars Support Airing Jerry Springer's Trashy Show," *Public Opinion* (Chambersburg, Pennsylvania), March 16, 1998, 13.
7. See Downey, *Closed Captioning*, 238–43.

8. Bill Stark, letter to Debra Sturdivant, Department of Education, January 13, 2000.
9. Bill Stark to Debra Sturdivant.
10. Bill Stark to Debra Sturdivant.
11. Bill Stark to Debra Sturdivant.
12. Bill Stark to Debra Sturdivant.
13. "Captioning Timeline Highlights," *Described and Captioned Media Program*, https://dcmp.org/learn/25-captioning-timeline-highlights.
14. "Video Streaming," Techopedia, last updated February 17, 2017, https://www.techopedia.com/definition/9927/video-streaming.
15. "The NAD Announces First-Ever Internet Video Streaming of Captioned Educational Videos," *Audiology Online*, December 23, 2002, https://www.audiologyonline.com/releases/nad-announces-first-ever-internet-5953.
16. Jennifer DiLorenzo, a school psychologist at St. Mary's School for the Deaf in Buffalo, New York, found the use of captioned media materials productive with both younger and older students. Using videos and DVDs with younger students was difficult at times because their language and reading levels tend to be delayed. When using this media, she took frequent breaks to discuss certain segments and ensure comprehension, and she always followed each showing with questions and activities to engrain the relevant information. As her students got older, she continued to use media as a stepping stone for learning. She described one particular lesson to show the impact the media had on her students' learning. Hundreds of other teachers have written detailed comments to Jason Stark in DCMP. For example, Jennifer DiLorenzo wrote, "It was a group of juniors and seniors, between 17 and 20 years old. Our topic was focused on the impacts of drinking alcohol. When we began, most of the students stated that they had 'heard it all before' and understood the dangers of drinking, as well as the concepts of alcohol. Prior to starting this lesson, I searched through the...titles and checked out a few that pertained to my overall objectives. After previewing them, I picked one that was perfect for my 'very in the know' students. I asked them to take note of any information presented that they were not previously aware of. As I sat and observed them watching this video, I could not help but smile.

    "It was clear to me that much of the information being shown was uncharted territory for most of them. When the video was complete, I asked them their thoughts. Most of them could not contain themselves, and the dialogue began to flow. They were shocked at how much they had not known and how well they could relate to the young adults in the video. That particular video sparked a discussion that continued for two additional class sessions. I could never have generated that kind of enthusiasm or thirst for more knowledge if I had tried to present that same information in a lecture format. Integrating captioned media into my lessons allows me to constantly present up-to-date and precise information that otherwise children who are deaf might miss." Jennifer DiLorenzo, "Youth and Captioned Media," 2003, *Described and Captioned Media Program*, https://dcmp.org/learn/156-youth-and-captioned-media.
17. Levesque, "Deaf People Are Patient People."
18. Audio Description Resource Guide. https://www.loc.gov/nls/about/services/reference-publications/guides/audio-description-resource-guide/.
19. Bill Stark, "Described and Captioned Media Program: Seventy Years of Progress

(1946–2016)," Described and Captioned Media Program, https://dcmp.org/learn/379-described-and-captioned-media-program-seventy-years-of-progress.

20. Patrick Loftus, "Which Students Find Closed Captions Helpful the Most?" 3Play Media/Oregon State University, December 9, 2016, updated June 3, 2019, https://www.3playmedia.com/2016/12/09/which-students-find-closed-captions-helpful-the-most/.
21. A. Evmenova, "Lights. Camera. Captions: The Effects of Picture and/or Word Captioning Adaptations, Alternative Narration, and Interactive Features on Video Comprehension by Students with Intellectual Disabilities" (PhD thesis, George Mason University, 2008); A. Evmenova and M. Behrmann, "Enabling Access and Enhancing Comprehension of Video Content for Postsecondary Students with Intellectual Disability," *Education and Training in Autism and Developmental Disabilities* 49, no. 1 (2014): 45–59.
22. V. Knight, B. R. McKissick, and A. Saunders, "A Review of Technology-Based Interventions to Teach Academic Skills to Students with Autism Spectrum Disorder," *Journal of Autism and Developmental Disorders* 43, no. 11 (2013): 2628–2648, https://doi.org/10.1007/s10803-013-1814-y; and K. A. Reagon, T. S. Higbee, and K. Endicott, "Using Video Instruction Procedures with and without Embedded Text to Teach Object Labeling to Preschoolers with Autism: A Preliminary Investigation," *Journal of Special Education Technology* 22, no. 1 (2007): 13–20.
23. J. L. Alty, A. Al-Sharrah, and N. Beacham, "When Humans Form Media and Media Form Humans: An Experimental Study Examining the Effects Different Digital Media Have on the Learning Outcomes of Students Who Have Different Learning Styles," *Interacting with Computers* 18, no. 5 (2006): 891–909; N. A. Beacham and J. L. Alty, "An Investigation into the Effects That Digital Media Can Have on the Learning Outcomes of Individuals Who Have Dyslexia," *Computers & Education* 47, no. 1 (2006): 74–93.
24. US Department of Labor, Office of Training and Education, "Presenting Effective Presentations with Visual Aids, May 1996, http://www.rufwork.com/110/mats/oshaVisualAids.html.
25. Since 2001, formal boards had been mandated by the US Department of Education for all major grants/cooperative agreements.
26. William Daugherty, letter to Jason Stark, January 29, 2016.
27. "National Black Deaf Advocates," https://www.nbda.org.
28. "Description Key," *Described and Captioned Media Program,* https://dcmp.org/learn/captioningkey/624.
29. Susan LaVenture, executive director of the National Association of Parents of Children with Visual Impairments (NAPVI), letter to Jason Stark, February 10, 2016.
30. "Description Tip Sheet," *Described and Captioned Media Program*, https://dcmp.org/learn/227.
31. "Description Tip Sheet."
32. Cecilia Robinson, letter to Jason Stark, February 2, 2016.
33. As DCMP's efforts with audio description developed, the NAD's program was commended by many more people who collaborated in providing the new resources, as well as blind and visually impaired users of the films. Ron Johnson, a national certified counselor who became legally blind as an adult and then returned to college to receive a MS in counseling, wrote, "These types of educational materials not only enabled me to successfully complete my education but they substantially enriched

my life in the process. As someone affected with a visual impairment, the enhanced materials made learning less tedious and more interesting. As a result, I can now serve others through counseling."

Support for DCMP's continued leadership similarly came from many other companies, including the National Center on Educational Materials for Learning. Many organizations recognized that the NAD and DCMP have excelled in providing this unique national educational accessible media resource. They have provided the only collection of free on-demand accessible media, led in the development of technologies to improve the creation and delivery of this media, and are viewed as a premier resource for educational description and captioning information. The Hearing Loss Association of America (HLAA) holds annual conventions advocates for the rights of hard of hearing people, conducts educational webinars, and has an extensive network of chapters and state organizations across the country. Lise Hamlin, director of Public Policy at HLAA, wrote, "Our members and constituents are avid users of captioning in general and captioned educational media in particular." Lisa Hamlin, letter to Jason Stark, February 10, 2016.

34. Ignacio Ponce, Chair, School of Modern Languages at Santa Barbara City College, letter to Jason Stark, January 21, 2016.
35. Lucia Hernandez, letter to Jason Stark, January 25, 2016.
36. Max Friedman, "Act Passed by Congress: Report on Conference," *Empire State News* 23, no. 1 (March 1960), Empire State Association of the Deaf, Inc.
37. Judith Holt, email to Bill Stark, November 9, 2005.
38. Bill Stark, email to Ernest Hairston, November 9, 2005.
39. Stark to Hairston.
40. Stark to Hairston.
41. "Film Fare," *The Silent Worker*, January 1963, 23.

### Chapter 15

1. "2008 NAD Conference Priorities," National Association of the Deaf, updated July 5, 2010, https://www.nad.org/about-us/priorities/2008-conference-priorities/.
2. Howard Fischer, "Harkins to Equip Cinemas for Deaf, Blind to Settle Arizona Suit," *East Valley Tribune* (Scottsdale, Arizona), October 13, 2011, https://www.eastvalleytribune.com/arizona/harkins-to-equip-cinemas-for-deaf-blind-to-settle-arizona-suit/article_cb9c88ec-f5fe-11e0-a0fb-001cc4c002e0.html.
3. John Waldo, "Movie Theater Captioning, Part 2 of 4: The Tide Turns," Deaf and Hard of Hearing Bar Association, April 18, 2018, http://www.deafbar.org/movie-theater-captioning-part-2-of-4-the-tide-turns/.
4. Waldo, "Movie Theater Captioning, Part 2 of 4."
5. Waldo, "Movie Theater Captioning, Part 2 of 4."
6. Waldo, "Movie Theater Captioning, Part 2 of 4."
7. John Waldo, "Movie Theater Captioning, Part 3 of 4: A National Requirement," Deaf and Hard of Hearing Bar Association, April 25, 2018, http://www.deafbar.org/movie-theater-captioning-part-3-of-4-a-national-requirement/.
8. Waldo, "Movie Theater Captioning, Part 3 of 4."
9. Department of Justice, Civil Rights Division, "Questions and Answers about the Department of Justice's Final Rule Under Title III of the ADA Requiring Movie Theaters to Provide Closed Movie Captioning and Audio Description," https://www.

ada.gov/regs2016/movie_captioning_qa.html.

10. Department of Justice, 28 CFR Part 36, CRT Docket no. 126; AG Order No. RIN 1190-AA63. Nondiscrimination on the Basis of Disability by Public Accommodations, Movie Theaters; Movie Captioning and Audio Description, https://www.ada.gov/regs2016/movie_rule.htm.
11. Waldo, "Movie Theater Captioning, Part 3 of 4."
12. John Waldo, "Movie Theater Captioning, Part 4 of 4: We're Not Done Yet," Deaf and Hard of Hearing Bar Association, May 2, 2018, http://www.deafbar.org/movie-theater-captioning-part-4-of-4-were-not-done-yet/.
13. Samantha Sauld, "How Many People Use Subtitles? Not Just the Deaf or Hard of Hearing" August 31, 2020, https://www.3playmedia.com/2015/08/28/who-uses-closed-captions-not-just-the-deaf-or-hard-of-hearing/.
14. Sveta Kouznetsova, "2018 Movie Theater Captioning Survey" Audio Accessibility, https://audio-accessibility.com/news/2018/04/movie-theater-captioning-access-survey-results/.
15. Kouznetsova, "2018 Movie Theater."
16. Kouznetsova, "2018 Movie Theater."
17. "Captioning and Subtitles." Virginia Tech. https://ssd.vt.edu/accommodations/deaf_hard_hearing/captioning-and-subtitles.html.
18. Katherine Bouton, "Why Movies Are So Frustrating for the Hard of Hearing," January 19, 2016, https://www.aarp.org/health/conditions-treatments/blogs/info-2016/movies-captions-deaf-hearing-loss.html.
19. The display came with a privacy visor so that other moviegoers could not see the captions. The closed captions were transmitted to the receivers over wireless frequencies. Each receiver consisted of a small organic light emitting diode display.
20. Gary Kerridge, "CaptiView: A Raw Deal for Deaf Cinema Goers," ABC (Australian Broadcasting Corp.), January 9, 2013, http://www.abc.net.au/rampup/articles/2013/01/09/3666873.htm.
21. Jason Om, "Deaf People Missing Out on Cinematic Experience Due to 'Appalling' Captioning Service." June 2, 2015. https://www.abc.net.au/news/2015-06-02/deaf-people-critical-of-movie-captioning-service/6512914.
22. Jackie Andre, "Please Fix the Malfunctioning Captiview Systems in Your Theaters." https://www.change.org/p/cineplex-please-fix-the-malfunctioning-captiview-systems-in-your-theatres.
23. Andre, "Please Fix."
24. "Commission Submits Letter of Support for Open Movie Captioning Act," *District of Columbia Bulletin*, November 27, 2018, https://content.govdelivery.com/bulletins/gd/DCWASH-21e1f27.
25. "Commission Submits Letter."
26. Claire Boston, "Glasses Show Captions for Deaf, Hard of Hearing," *Washington Times*, June 16, 2014, https://www.washingtontimes.com/news/2014/jun/16/glasses-show-captions-for-deaf-hard-of-hearing/.
27. Boston, "Glasses Show Captions."
28. James Jones, email to author, August 24, 2018.
29. Rachel Kurzius, "Movie Theaters Are Often Frustrating for the Deaf Commu-

nity. This Bill Could Help," *DCist.com*, September 19, 2018, https://dcist.com/story/18/09/19/open-captions-bill-dc/.
30. Bennett J. Loudon, "Feds Plan New Law on Movie Captions," *Democrat and Chronicle* (Rochester, New York), December 3, 2014, A1.
31. Loudon, "Feds Plan New Law," A7.
32. "SouthSide Works Cinema Now Offers Open Captioning," National Association of the Deaf, October 18, 2018, https://www.nad.org/2018/10/18/southside-works-cinema-now-offers-open-captioning/.
33. Waldo, "Movie Theater Captioning, Part 4 of 4."
34. Joe Mullin, "9th Circuit Rules Netflix Isn't Subject to Disability Law. Are Websites 'Public Accommodations'? Federal Courts Have Split on the Question," *Ars Technica*, April 2, 2019, https://arstechnica.com/tech-policy/2015/04/9th-circuit-rules-netflix-isnt-subject-to-disability-law/.
35. Jon Christian, "How Netflix Alienated and Insulted Its Deaf Subscribers," *The Week*, January 30, 2014, https://theweek.com/articles/452181/how-netflix-alienated-insulted-deaf-subscribers.
36. Jamie Pauls, "Audio Description Comes to Netflix," American Federation of the Blind, July 2015, https://www.afb.org/aw/16/7/15436.
37. "National Association of the Deaf Announces Landmark Settlement with Harvard to Improve Online Accessibility," Disability Rights Education & Defense Fund, November 27, 2019, https://dredf.org/2019/11/27/landmark-settlement-with-harvard-to-improve-online-accessibility/.
38. "National Association of the Deaf Announces."
39. "Ask Howard Anything," National Association of the Deaf, March 2019, https://silentgrapevine.com/2019/03/ask-howard-anything-march-2019.html.
40. "Mission Statement," TDI Online, http://tdi-online.org.
41. "In the Matter of Implementation of Section 255 of the Telecommunications Act of 1996, Access to Telecommunications Services, Telecommunications Equipment, and Customer Premises Equipment by Persons with Disabilities, WT Docket No. 96-198," FCC.gov, August 14, 1998, https://transition.fcc.gov/Bureaus/Wireless/Comments/fcc98055/td-r.pdf.
42. "Remarks of FCC Chairman Tom Wheeler as Prepared for Delivery, TDI Conference, Baltimore, MD, August 20, 2015," FCC.gov, DOC-334979A1.pdf.
43. "A Path toward an Accessible World: 1968–2018," A TDI World Anniversary Commemorative Edition, TDI, 49, no. 2 (2018), https://tdiforaccess.org/wp-content/uploads/2019/09/TDI-World-49_2-50th-Anniversary-Commemorative-edition.pdf.
44. "A Path toward an Accessible World," 71.
45. "A Path toward an Accessible World," 84.
46. "A Path toward an Accessible World," 3.
47. Catalin Cimpanu, "Yahoo Groups to Shut Down for Good on December 15, 2020," ZDNet, October 12, 2020, https://www.zdnet.com/article/yahoo-groups-to-shut-down-for-good-on-december-15-2020/.
48. CC Crusaders Facebook page (@CCCrusaders), https://www.facebook.com/CCCrusaders/.
49. Captioning Activism and Community (CCAC), http://CCACaptioning.org.

50. "Cinema Advocacy," Captioning Activism and Community, November 2012, http://ccacaptioning.org/show-captions-cinema-captioning-advocacy-november-2/.

CHAPTER 16

1. Work at the local residential school for deaf students continued as Jason Stark assumed the role of technician's assistant, working on campus to repair, clean, and upgrade instructional technology on campus. This work sparked his overall love of technology and his desire to find a career that helped people. His experiences also included the administration of staff in fifty-eight media libraries (depositories) across the United States. His two degrees from the University of South Carolina, in business administration and information systems management, had provided him with the knowledge and skills for the development of innovative technology leadership. This would not only result in government savings of millions of dollars over the past twenty-five years, but it also brought significant improvements in multimedia access and expansion of services to children with disabilities in the US. Jason also had ongoing responsibility for the financial and operations management of the federal grants with budgets up to $17 million.
2. Courtney Duerig, "YouTube's Automatic Captions Prove Insufficient for ADA Compliance," February 24, 2016, https://www.automaticsync.com/captionsync/youtube-automatic-captions/.
3. Rikki Poynter created the #NoMoreCraptions campaign to address this: https://mashable.com/2016/11/11/youtube-closed-captioning-nomorecraptions/. People actively use the hashtag on Twitter: https://twitter.com/hashtag/nomorecraptions?f=live.
4. Emma Grey Ellis, "The Problem With YouTube's Terrible Closed 'Craptions,'" *Wired*, January 18, 2019, https://www.wired.com/story/problem-with-youtubes-terrible-closed-craptions/.
5. For some of the many resources available through DCMP, see https://dcmp.org/learn/static-assets/594/literacy-and-reading-8-2020.pdf.
6. "How to View DCMP Media on Roku," *Described and Captioned Media Program*, https://dcmp.org/learn/392-how-to-view-dcmp-media-on-roku.
7. When JoAnn McCann retired in 2018, Glinda Hill replaced her as DCMP project officer. In 2011, Ernie Hairston retired from the Department of Education. Bill Stark moved to a half-time position in DCMP and into semiretirement. Bill had been the project director for more than twenty years. Before retiring in 2017 he served as communications and accountability administrator for the program, focusing on client communications and working with the program staff on a broad range of client recruitment and promotional activities. During his tenure as project director, millions of children and adults had been provided opportunities for cultural experiences and learning through the thousands of educational media items that the project made accessible or produced. A crowning achievement in 2006 was Bill's presiding over the dramatic and hugely important addition of project services to blind and visually impaired students. All media henceforth to be acquired and distributed by the project is to have audio description along with captions.
8. Karen Peltz Strauss oversaw the drafting of the closed captioning quality standards

as deputy bureau chief of the FCC's Consumer and Governmental Affairs Bureau. She did this under FCC chairman Tom Wheeler. It took years and extensive negotiations between consumers and industry to make these happen.

9. See "In the Matter of Closed Captioning of Video Programming, Telecommunications for the Deaf and Hard of Hearing, Inc., Petition for Rulemaking, CG Docket No. 05-231" FCC.gov, https://docs.fcc.gov/public/attachments/FCC-14-12A1.pdf.
10. Bill Stark, "Read Captions Across America," *Described and Captioned Media Program,* https://dcmp.org/learn/154-read-captions-across-america.
11. Donna Hom, letter to Jason Stark, February 3, 2016.
12. Through the years, many technologies evolved in the program. For example, whereas RealPlayer was first used to stream videos, the installation of Adobe Flash Player allowed the videos to be viewed in any browser.
13. "What Is Universal Design," Centre for Excellence in Universal Design, http://universaldesign.ie/What-is-Universal-Design/.
14. Bill Stark, "Described and Captioned Media Program: Seventy Years of Progress (1946–2016)," *Described and Captioned Media Program,* https://dcmp.org/learn/379-described-and-captioned-media-program-seventy-years-of-progress.
15. Stark, "DCMP's Keys to Access."
16. Linda Cannon, letter to Jason Stark, January 27, 2016.
17. "Educational Media Producer and Accessibility Survey Results," *Described and Captioned Media Program,* https://dcmp.org/learn/226-educational-media-producer-and-accessibility-survey-results.
18. Bill Stark, email to author, September 22, 1918.
19. See Blake E. Reid, *Third Party Captioning and Copyright: A G3ict Policy White Paper* (unpublished paper).
20. "eLearning Resources," *Described and Captioned Media Program,* https://dcmp.org/elearning.
21. "Help Center: Quick Classes," *Described and Captioned Media Program,* https://dcmp.org/learn/398-quick-classes.

## Epilogue

1. E. Alex Jung and Alison Willmore, "A Guide to Director Bong Joon Ho's Deeper Cuts," *New York Vulture,* February 11, 2020, https://www.vulture.com/2020/02/best-bong-joon-ho-movies-parasite-director-must-sees.html.
2. John A. Gough, "Gullible's Travels" (unpublished personal memoirs of John A. Gough), shared by Susan Gough Updegraff. January 2019, 88.
3. Jesse Marpoe, "*Children of a Lesser God* (1986)," *Just Another Weblog* (blog), November 14, 2015, https://sites.psu.edu/jessemarpoe5992/2015/11/14/children-of-a-lesser-god-1986/.
4. Emanuel Levy, "Reel/Real Impact: *Children of a Lesser God* (1986)," February 20, 2006, http://emanuellevy.com/comment/children-of-a-lesser-god-1986-cultural-impact-4/.
5. Billing Code: 4410-13, Department of Justice, 28 CFR Part 36, CRT Docket No. 126; AG Order No. RIN 1190-AA63, Nondiscrimination on the Basis of Disability by Public Accommodations, Movie Theaters; Movie Captioning and Audio Descrip-

tion, https://www.ada.gov/regs2016/movie_rule.htm.

6. "About DCMP," *Described and Captioned Media Program*, https://dcmp.org/about-dcmp?A=2.
7. Dan Gingiss, "How Social Media Became a Game-Changer for the Deaf Community," Social Media Today, January 9, 2018, https://www.socialmediatoday.com/news/how-social-media-became-a-game-changer-for-the-deaf-community/514223/.
8. Gingiss, "How Social Media Became a Game-Changer."
9. Gingiss, "How Social Media Became a Game-Changer."
10. Tania Pierre, "How Social Media Empowers and Connects People—D/deaf and Hearing." Deaf Unity, April 10, 2019, https://deafunity.org/article_interview/how-social-media-empowers-and-connects-people-deaf-and-hearing/.
11. Pierre, "How Social Media Empowers."
12. Pierre, "How Social Media Empowers."
13. I. Kožuh and M. Debevc, "Challenges in Social Media Use Among Deaf and Hard of Hearing People," in *Social Networks Science: Design, Implementation, Security, and Challenges*, eds. N. Dey, R. Babo, A. Ashour, V. Bhatnagar, and M. Bouhlel (Cham: Springer, 2018), https://link.springer.com/chapter/10.1007/978-3-319-90059-9_8.
14. Sofia Enamorado, "Social Video and the Changing Trends in Video Accessibility, 3Play Media, December 9, 2019, https://learn.3playmedia.com/blog-3play-media/social-video-and-the-changing-trends-in-video-accessibility.
15. Enamorado, "Social Video."
16. Enamorado, "Social Video."
17. Amma Marfo, "How to Caption Your Content Across All Social Media Platforms," Social Media Week, April 16, 2019, https://socialmediaweek.org/blog/2019/04/how-to-caption-your-content-across-all-social-media-platforms/.
18. Marfo, "How to Caption."
19. Marfo, "How to Caption."
20. @Kelly4Access, Instagram, December 15, 2020, https://www.instagram.com/tv/CI2A5LQFszg/?igshid=qda2ey6in39z.
21. Gael Hannan, "We Want Captions," Hearing Health Matters, November 25, 2020, https://hearinghealthmatters.org/betterhearingconsumer/2020/we-want-captions/.
22. Hannan, "We Want Captions."
23. Shari Eberts, "Two Down! Only Zoom to Go" (petition), Change.org, May 31, 2020, https://www.change.org/p/zoom-provide-free-captions-for-people-with-hearing-loss-on-video-conferencing-platforms/u/26778357.
24. Eberts, "Two Down!"
25. Shari Eberts, "People with Hearing Loss Shouldn't Have to Pay Zoom for Captions," *Washington Post*, December 17, 2020, https://www.washingtonpost.com/opinions/2020/12/17/zoom-captions-hearing-loss-free/.
26. Association of Late-Deafened Adults (@ALDA_Inc), Twitter, November 17, 2020, https://twitter.com/ALDA_Inc/status/1328825761164189697?s=20.
27. Liam O'Dell, "Social Media is Embracing Audio More Than Ever—But Deaf Users Are Being Left Behind," TechRadar, December 2020, https://www.techradar.com/news/social-media-is-embracing-audio-more-than-ever-but-deaf-users-are-being-left-behind.
28. O'Dell, "Social Media is Embracing Audio."

29. Veronica Figueroa Fernandez, "6 Tips on Making Your Content Accessible to the Deaf and HOH Community." Laughing Samurai, September 28, 2020, https://laughingsamurai.com/social-media-marketing/making-your-content-accessible-to-the-deaf-community/.
30. Fernandez, "6 Tips."
31. T.J. McCue, "Verizon Media Says 69 Percent of Consumers Watching Video with Sound Off," *Forbes*, July 31, 2019, https://www.forbes.com/sites/tjmccue/2019/07/31/verizon-media-says-69-percent-of-consumers-watching-video-with-sound-off/?sh=5dbd3b0b35d8.
32. Joan Haviland, email to author, January 17, 2019.
33. Sean Forbes, "No Captions Again," Facebook, January 1, 2019, https://www.facebook.com/watch/?v=405227030019428.
34. Rachel Treisman, "The World's Largest Movie Theater Chain Is Adding Open Captions at 240 U.S. Locations." *National Public Radio*, October 21, 2021, https://www.npr.org/2021/10/21/1048017853/amc-theatres-open-captioning-movie-accessibility.
35. Forbes, "No Captions Again."

### Afterword

1. "There's a Way to Do It Better—Find It. Thomas Edison? David Sarnoff? Apocryphal?" (Online). Available https://www.brainyquote.com/authors/thomas_a_edison
2. "37 Quotes from Thomas Edison That Will Inspire Success" (Online). Available https://www.inc.com/kevin-daum/37-quotes-from-thomas-edison-that-will-bring-out-your-best.html.
3. Malcolm J. Norwood, "Don't Scramble the Wrong Egg," presented at the National Technical Institute for the Deaf at Rochester Institute of Technology, 1980. https://dcmp.org/learn/83.
4. Norwood, "Don't Scramble."
5. B. Braverman and B. J. Cronin, *Captioning: Shared Perspectives.* Rochester, NY. National Technical Institute for the Deaf, 1980. https://dcmp.org/learn/static-assets/nadh83.pdf.
6. Julie Moran Alterio, "Technology Makes Webcasts Accessible to the Deaf." *News-Journal* (Mansfield, Richland, Ohio, October 24, 2005), 16.
7. "How Bird Hunting in North Carolina Saved Captioning". (Online) https://dcmp.org/learn/36-how-bird-hunting-in-north-carolina-saved-captioning. 2004.

### Acknowledgments

1. Carol Eileen (Evanhoe) Sponable, "The Oklahoma School of [sic] the Deaf: The Supt. John A. Gough Era." *OKCAD Newsletter*, September, 1996.
2. The nickname was given to Bill by Gerald Burstein, a deaf film depository manager from California and a Certified Professional Parliamentarian, and that honor stayed with Bill through the years as he transitioned CFD and modernized resources that included captioning in Spanish. He initiated the process of description in the program to make films also accessible to blind and deaf-blind learners. Bill deserves much credit for the positive impact of captioning on the language skills of deaf students today through both their primary and vicarious learning experiences. His devotion to building a major online resource is commendable. It is widely used today

by educators and parents of deaf children.

3. Robert R. Davila to Harry G. Lang, September 5, 2018.
4. Jeff Hutchins was founder in 1986, with Dave Crane, of VITAC, which became the largest captioning company in the country.

# INDEX

## ABOUT THE AUTHOR

Harry G. Lang is professor emeritus at the National Technical Institute for the Deaf at Rochester Institute of Technology, where he taught for forty-one years. Lang attended the Western Pennsylvania School for the Deaf in Pittsburgh, received a bachelor's degree from Bethany College in West Virginia and a master's degree at Rochester Institute of Technology, and earned an EdD from the University of Rochester. He received an honorary doctorate from Gallaudet University in 2013.

Dr. Lang has published nine books and numerous book chapters and articles, including: *Silence of the Spheres: The Deaf Experience in the History of Science* (1994), *Deaf Persons in the Arts and Sciences: A Biographical Dictionary* (with Bonnie Meath-Lang, 1995), *A Phone of Our Own: The Deaf Insurrection against Ma Bell* (2000), *Educating Deaf Students: From Research to Practice* (with Marc Marschark, and John Albertini 2002), *Edmund Booth: Deaf Pioneer* (2004), *Teaching from the Heart and Soul: The Robert F. Panara Story* (2007), *Moments of Truth: Robert R. Davila, The Story of a Deaf Leader* (with Oscar Cohen and Joseph Fischgrund, 2008), and *Fighting in the Shadows: Untold Stories of Deaf People in the Civil War (2017).*

Among his many awards are TDI's James C. Marsters Promotion Award "for outstanding contributions to improving the visual accessibility of information, entertainment or telecommunications in the United States" (2001). Dr. Lang is the second recipient of the Dr. John S. and Dr. Betty J. Schuchman Deaf History Award from Gallaudet University. First established in 2014, this award is given every five years to recognize significant scholarship in the field of Deaf history.

NATIONAL
ENDOWMENT
FOR THE
HUMANITIES

*Turn on the Words! Deaf Audiences, Captions, and the Long Struggle for Access* by Harry G. Lang has been made possible in part by the National Endowment for the Humanities: Exploring the human endeavor (www.neh.gov).

Any views, findings, conclusions, or recommendations expressed in this book do not necessarily represent those of the National Endowment for the Humanities.